VITAMIN C

Edited by **Amal H. Hamza**

Vitamin C
http://dx.doi.org/10.5772/66058
Edited by Amal H. Hamza

Contributors

Domenico Mastrangelo, Fadime Eryılmaz Pehlivan, Sudha J Devaki, Reshma Laly Ravindran, Mahendra Nath Roy, Subhadeep Saha, Tetsuo Yamamoto, Manabu Kinoshita, Carmen García-Rodríguez, Alejandro Gordillo-García, Mario Altamirano-Lozano, Francisco Nualart, Luciano Ferrada, Katterine Salazar

Published by InTech

Janeza Trdine 9, 51000 Rijeka, Croatia

Publishing Process Manager Edi Lipovic
Technical Editor SPi Global
Cover InTech Design team

First published July, 2017
Printed in Croatia
Legal deposit, Croatia: National and University Library in Zagreb

Additional hard copies can be obtained from orders@intechopen.com

Vitamin C, Edited by Amal H. Hamza
p. cm.
Print ISBN 978-953-51-3421-3
Online ISBN 978-953-51-3422-0

For EU product safety concerns:
IN TECH d.o.o., Prolaz Marije Krucifikse Kozulić 3, 51000 Rijeka, Croatia,
info@intechopen.com or visit our website at intechopen.com.

Meet the editor

Dr. Amal Hamza is an associate professor at the Biochemistry Department, King Abdulaziz University (KAU), Saudi Arabia, since 2011. Previously, he worked as a teaching assistant and associate professor at the Department of Biochemistry and Nutrition, Ain Shams University, Egypt.
He has over 20 years of teaching and research experience. He has developed and taught undergraduate classes in biochemistry and nutrition and served on thesis and dissertation committees. He has coauthored more than 30 publications and cosupervised many undergraduates and masters' degree students. His research interest focused on studying the molecular and biochemical mechanism of natural products and essential nutrients in managing and ameliorating several diseases.

Contents

Preface

This book has been devised to offer an overview of currently hot topics related to vitamin C, which will be very important to people who are interested in biochemistry and biology.

Although vitamin C has been discovered long ago, research into vitamin C still progresses rapidly. Vitamin C is a simple molecule with a powerful effect. We decided to introduce this book to highlight the new advances and research related to this very important vitamin.

The authors were selected according to their prospective contribution to the scientific field. Authors were asked to submit a chapter related to the topic. All submitted chapters were subjected to a referring procedure including plagiarism check.

The chapters of this book include basic information about vitamin C function, sources, and analysis. Radioprotective and antioxidant effect of vitamin C is also considered in this book. Also, the anticarcinogenic effect of vitamin C is introduced. Furthermore, we considered the encapsulation technique used in vitamin C preparation. Finally, recent advances in vitamin C transporter are illustrated.

We hope that the book's readers will find interesting topics that will greatly help support the scientific research on this powerful vitamin.

Amal H. Hamza
King Abdulaziz University, Jeddah, KSA
Ain Shams University, Cairo, Egypt

Section 1

What is Vitamin C

Chapter 1

Vitamin C: Sources, Functions, Sensing and Analysis

Sudha J. Devaki and Reshma Lali Raveendran

Additional information is available at the end of the chapter

http://dx.doi.org/10.5772/intechopen.70162

Abstract

Vitamin C is a water-soluble compound found in living organisms. It is an essential nutrient for various metabolism in our body and also serves as a reagent for the preparation of many materials in the pharmaceutical and food industry. In this perspective, this chapter can develop interest and curiosity among all practicing scientists and technologists by expounding the details of its sources, chemistry, multifunctional properties and applications.

Keywords: vitamin C, biomarker, antioxidant, sensors

1. Introduction

Vitamin C also known as ascorbic acid (AA) is an essential nutrient in many multicellular organisms, especially in humans. Ascorbic acid is a water-soluble vitamin and is found in variable quantities in fruits and vegetables and organ meats (e.g. liver and kidney). Deficiency of vitamin C causes scurvy, widespread connective tissue weakness and capillary fragility. Among chemists, it is used as a reagent for the preparation of fine chemicals, enzymatic reagent and nanomaterials. Consequently, the detection and quantification of ascorbic acid in food samples, products and nutraceuticals is receiving overwhelming importance among researchers, medical practitioners and also in the pharmaceutical and food industry. **Figure 1** shows the schematic representation of the sources and multifunctional applications of vitamin C in the metabolism of our body.

2. Sources

Vitamin C (**Figure 2**) is abundantly available in many natural sources, including fresh fruits and vegetables. The richest sources of ascorbic acid including Indian gooseberry, citrus fruits

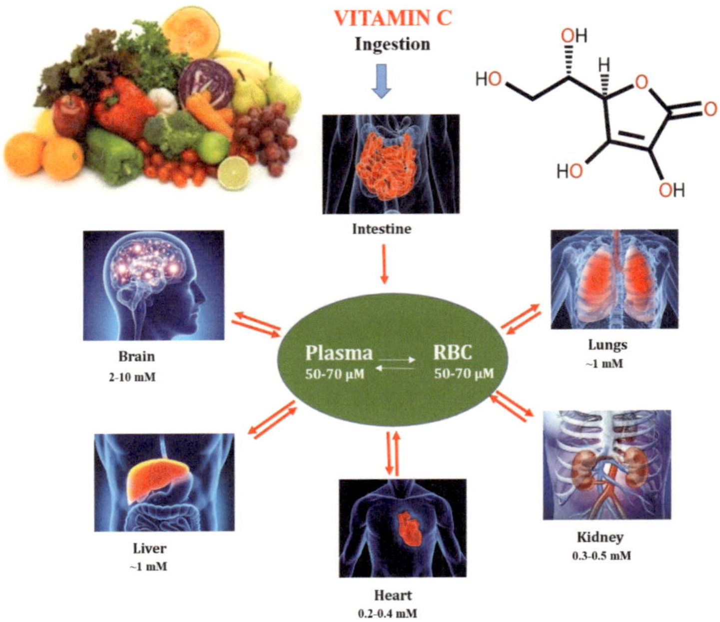

Figure 1. Sources and multifunctional role of vitamin C in metabolic processes in human body.

such as limes, oranges and lemons, tomatoes, potatoes, papaya, green and red peppers, kiwifruit, strawberries and cantaloupes, green leafy vegetables such as broccoli, fortified cereals and its juices are also rich sources of vitamin C.

Another source of vitamin C is animals. They usually synthesize their own vitamin C and are highly concentrated in the liver part [1, 2]. Sources of vitamin C and its content are given in **Figure 3**. Average daily recommended amounts of vitamin C for different ages are also given in **Table 1**.

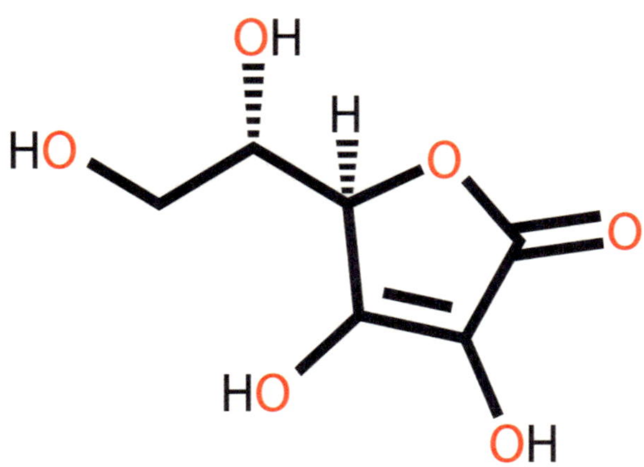

Figure 2. Structure of vitamin C.

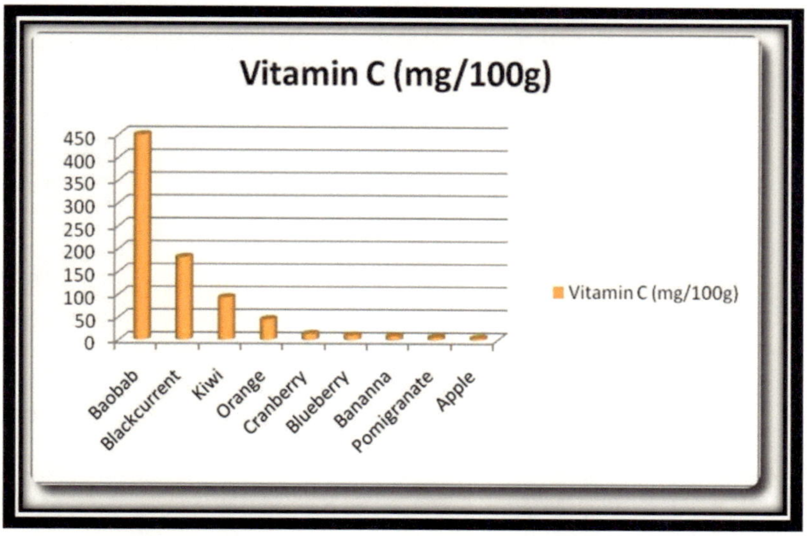

Figure 3. Amount of vitamin C in different sources.

Infants					
Age			**Adequate intake (AI) mg/day**		**Upper intake level (UL)**
0–6 months[a]			25		–
7–12 months[b]			30		–
Children and adolescents					
Age	EAR[c]		RDI[d]		
	mg/day		mg/day		mg/day
1–3	25		35		400
4–8	25		35		650
9–14	28		40		1200
14–18	28		40		1800
Adults					
Men			Women		Both
Age	EAR[c]	RDI[d]	EAR[c]	RDI[d]	
	mg/day	mg/day	mg/day	mg/day	mg/day
19–30	30	45	30	45	2000
31–50	30	45	30	45	2000
51–70	30	45	30	45	2000
>70	30	45	30	45	2000

Infants					
Age			Adequate intake (AI) mg/day		Upper intake level (UL)
Motherhood					
Pregnancy			Lactation		Both
Age	EAR[c]	RDI[d]	EAR[c]	RDI[d]	
	mg/day	mg/day	mg/day	mg/day	mg/day
14–18	38	55	58	80	2000
19–30	40	60	60	85	2000
31–50	40	60	60	85	2000

[a]Calculated as per the average intake of breast milk.
[b]Calculated on a body weight basis of infants.
[c]Estimated average requirement.
[d]Recommended dietary intake.

Table 1. Dietary intake of vitamin C [3–5].

3. Functions

Vitamin C plays an important role in many physiological processes in humans. It is needed for the repair of tissues in all parts of the body. The important functions of vitamin C include the formation of protein used to make skin, tendons, ligaments, and blood vessels for healing wounds and forming scar tissue, for repairing and maintaining cartilage, bones, and teeth and aid in the absorption of iron. It can also act as a reducing and capping agent for metal nanoparticles.

3.1. As reducing and capping agent

Ascorbic acid acts as a reducing and capping agent for the synthesis of metal nanoparticles such as silver, gold, copper, etc. Ascorbic acid molecules can cap or surround the particle and prevent the uncontrolled growth of the particles to micron-sized dimensions. A study by Khan et al., in 2016 reported the synthesis of copper nanoparticle by using ascorbic acid as both the reducing agent [6]. Sun et al. reported in *Journal of Materials Science* in 2009 that gold nanoparticles can be synthesised in reverse micelles without the addition or introduction of any other reducing or capping reagent [7]. In analytical methods in 2014, D'souza et al. reported the use of AA-Au NPs as a colorimetric probe for detection of dichlorvos in water and wheat samples. The influence of AA concentration on the aggregation induced by dichlorvos in AA-AuNPs (**Figure 4**) and the optical property of the AA-Au NPs was investigated by UV-Vis spectroscopy [8].

http://dx.doi.org/10.5772/intechopen.70162

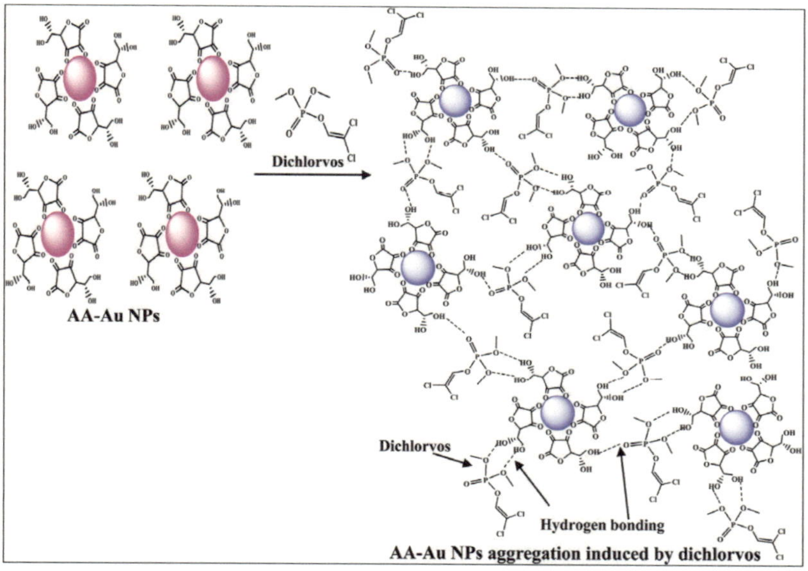

Figure 4. Analytical process for detecting dichlorvos using AA-Au NPs as a colorimetric probe [8].

3.2. Antioxidant activity

One of the important properties of vitamin C is its antioxidant activity. Antioxidant activity of vitamin C helps to prevent certain diseases such as cancer, cardiovascular diseases, common cold, age-related muscular degeneration and cataract.

3.3. In cancer treatment

Since 1970, it has been known that high dose of vitamin C has beneficial effects on the survival time in patients with terminal cancer, which was reported by Cameron, Campbell, and Pauling. Research is undergoing in detail for using vitamin C in cancer treatment [9–11]. One of the studies suggests that pharmacologic doses of vitamin C might show promising effects on the treatment of tumours [12]. Vitamin C can act as pro-oxidant and it can generate hydrogen peroxide [13, 14]. Administration of high dose of vitamin C gives long survival times for patients with advanced cancers.

The continuing attack of DNA by unquenched reactive oxygen species is believed to cause cancer. As a physiological antioxidant, ascorbic acid plays a role in the prevention of oxidative damage to DNA, which is elevated in cells at sites of chronic inflammation and in many pre-neoplastic lesions. Most DNA damage is repaired metabolically; however, the frequency of elevated steady-state levels of oxidized DNA bases is estimated to be sufficient to cause mutational events. The base damage product 8-hydroxy-2′-deoxyguanosine has been found to be elevated in individuals with severe vitamin C deficiency and to be reduced by supplementation with vitamins C and E [15].

Oxidative stress, arising as a result of an imbalance between free radical production and antioxidant defences, causes damage to a wide range of molecular species including lipids, proteins and nucleic acids. Oxidative stress caused by free radicals is shown in **Figure 5**. The best way to ensure adequate intake of the antioxidant nutrients, such as vitamins C and E and various kinds of minerals, is through a balanced diet consisting of 5–8 servings of fruits and vegetables per day.

3.4. In cardiovascular diseases

The antioxidant property of vitamin C helps for the treatment of cardiovascular diseases. Vitamin C has the capability for reducing monocyte adherence to the endothelium, improving endothelium-dependent nitric oxide production and vasodilation and reducing vascular smooth-muscle-cell apoptosis, which prevents plaque instability in atherosclerosis [15]. The oxidative damage including the oxidative modification of low-density lipoproteins is a major cause of cardiovascular disease. The antioxidant property of vitamin C helps to reduce this to a certain extent [16, 17].

3.5. In common cold

Pauling in 1970 suggested that vitamin C can be used for the treatment of common cold [18]. There are so many reports in Cochrane Database Syst. Review showing the use of prophylactic

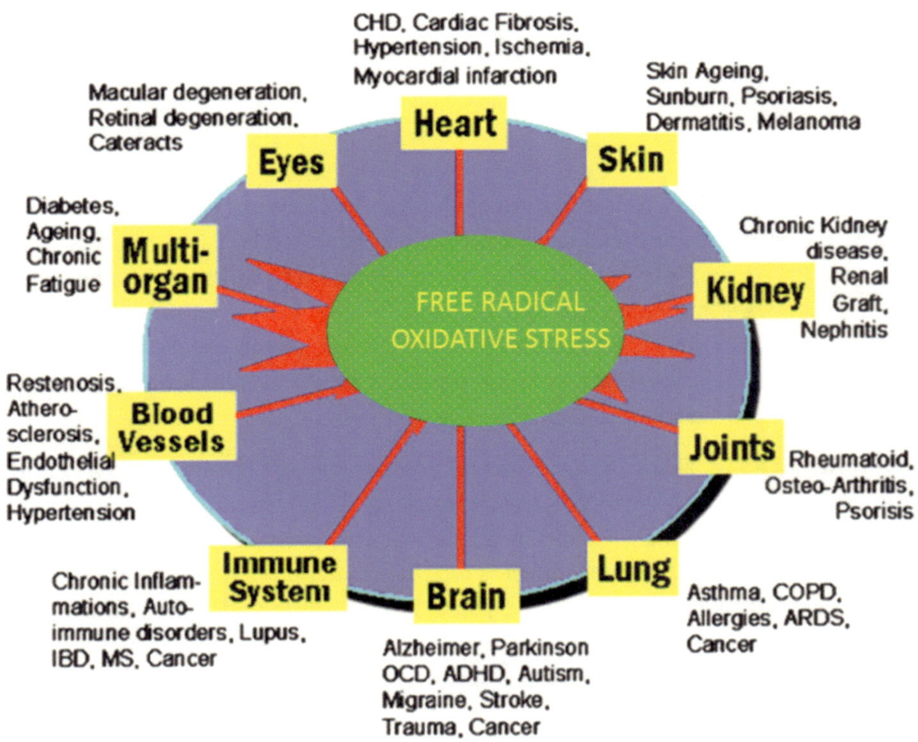

Figure 5. Oxidative stress by free radicals.

http://dx.doi.org/10.5772/intechopen.70162

vitamin C reduces the cold duration in adults and children [19]. The use of vitamin C might reduce the duration of common cold due to its anti-histamine effect of high dose of vitamin C [20]. However, the results are inconsistent and still research is undergoing in this field. There are Database Syst. Review showing the use of prophylactic vitamin C reduces the cold duration in adults and children [19]. The use of vitamin C might reduce the duration of common cold due to its anti-histamine effect of high dose of vitamin C [20].

3.6. In age-related macular degeneration (AMD) and cataract

Age-related macular degeneration (AMD) and cataracts are two of the main causes of vision loss in older. Oxidative stress might contribute to the aetiology of both conditions. Thus, researchers have taken interest in the role of vitamin C and other antioxidants in the development and treatment of these diseases. There are many reports to study the role of vitamin C in AMD and cataract [21–23]. Results from two studies indicate that vitamin C intakes greater than 300 mg/day reduce the risk of cataract formation by 75% [16, 24, 25].

All the studies indicate that the vitamin C formulations might slow AMD progression and reduce the high risk of developing advanced AMD. AMD is shown in **Figure 6**.

3.7. Antioxidant mechanism

Vitamins C can protect the body against the destructive effects of free radicals. Antioxidants neutralize free radicals by donating one of their own electrons, ending the electron-stealing reaction, as shown in **Figure 7**. The antioxidant nutrients themselves do not become free radicals by donating an electron because they are stable in either form or act as scavengers, helping to prevent cell and tissue damage that could lead to cellular damage and disease.

Ascorbic acid reacts with free radicals undergoing single-electron oxidation to produce a relatively poor reactive intermediate, the ascorbyl radical, which disproportionates to ascorbate and dehydroascorbate. Thus, ascorbic acid can reduce toxic, reactive oxygen species superoxide anion ($O_2^{\bullet}$) and hydroxyl radical ($OH^{\bullet}$), as well as organic (RO_2) and nitrogen ($NO_2^{\bullet}$) oxy

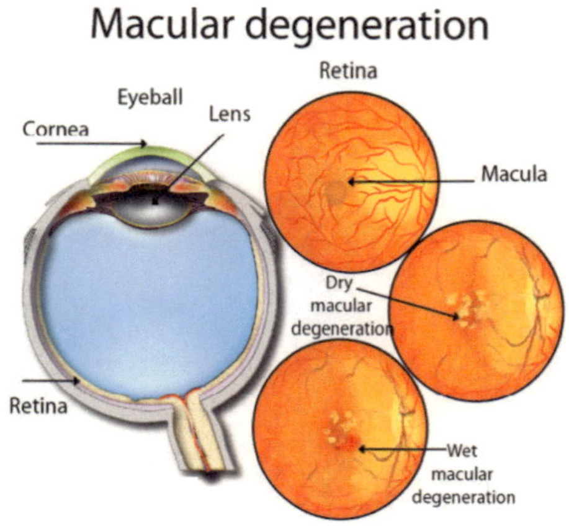

Figure 6. Age-related macular degeneration.

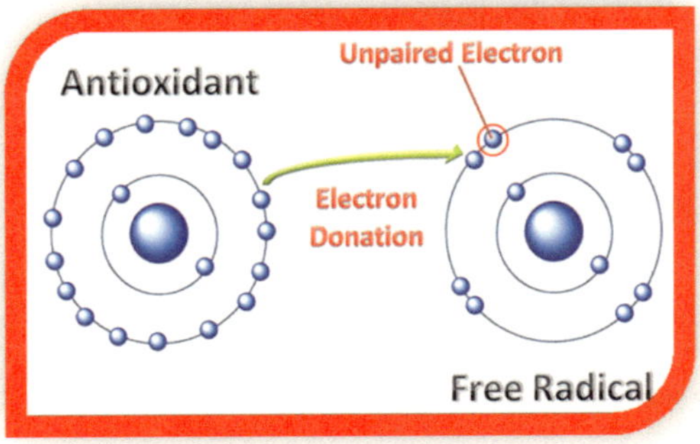

Figure 7. Antioxidant mechanism.

radicals. Those reactions are likely to be of fundamental importance in all aerobic cells. This reaction is the basis of most of the biological functions of ascorbic acid. The mechanism of free radical action on DNA is shown in **Figure 8**.

Vitamin C also has role in protecting other vitamins (vitamin A and vitamin E) from the harmful effects of oxidation. Vitamin C helps in protecting gums and retards ageing. It strengthens the general physical condition by removing toxic metals from the body. Vitamin C reduces the formation of cataract and hence useful in the treatment of glaucoma.

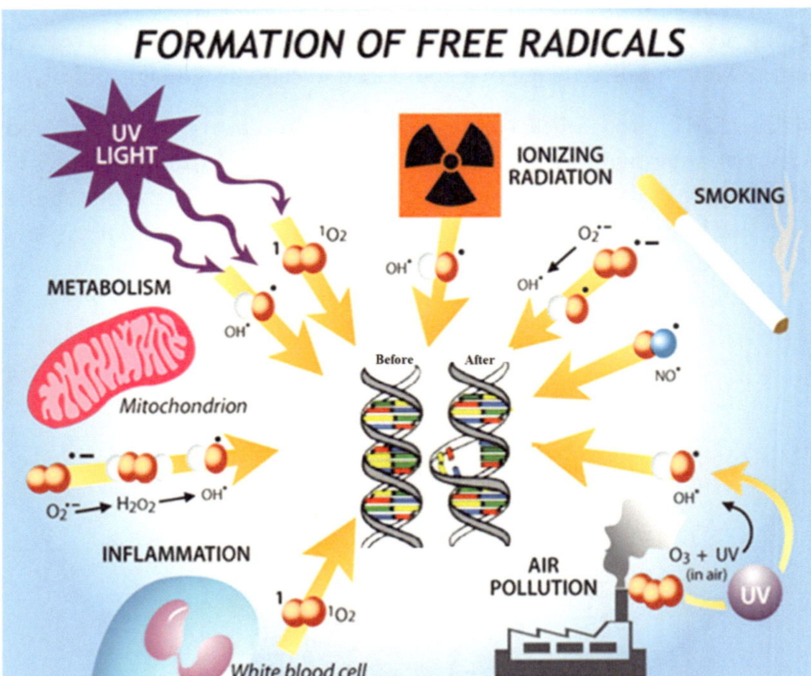

Figure 8. Mechanism of free radical action on DNA.

3.8. Synthesis of protein

Another important function of vitamin C is its role in the synthesis of protein. Vitamin C helps in the synthesis of collagen. Collagen protects our skin from wrinkling and makes our skin firm and strong. Collagen also protects and supports organs and other soft tissues. One of the amino acids used to build collagen—hydroxyproline—is only synthesized when vitamin C is available. Functions of vitamin C on skin.

3.9. Functions of vitamin C on skin

3.9.1. Photoprotection

Vitamin C reduces the damage caused by UV-light exposure. It cannot act as a sunscreen since it cannot absorb UV light. But the antioxidant activity of vitamin C helps to protect UV damage caused by free radicals [26]. In response to UV light, vitamin C transport proteins are increased, suggesting an increased need for vitamin C uptake for adequate protection [27, 28]. Addition of keratinocytes on vitamin C reduces damages caused by UV light and lipid peroxidation, limits the release of pro-inflammatory cytokines and protects against apoptosis [29]. Many studies have suggested that vitamin C consumption alone will not reduce the effect of UV exposure; however, a combination of vitamin C and E effectively increases minimal erythemal dose (MED) (the lowest dose of ultraviolet radiation (UVR) that will produce a detectable erythema 24 hours after UVR exposure) and decreases erythema-induced blood flow to damaged areas of skin [30]. Thus, interactions between the two antioxidant vitamins may be necessary to achieve UV protection. The topical application of vitamin C also reduces the effect of UV exposure and skin wrinkling and skin tumour [31]. Vitamin C reduced the number of sunburned cells, decreased erythema response and reduced DNA damage induced by UV exposure [27]. The combination of antioxidant vitamins decreased the immunosuppressive effects of UV exposure, increased MED and decreased cell damage [32, 33].

3.9.2. Wound healing

Vitamin C plays a key role in healing wound by the formation of collagen, connective tissue [34–36]. The new tissue is rebuilt with the help of collagen framework. This function is supported by its co-factor vitamin C. Besides this, vitamin C performs as a strong antioxidant and immune system modulator [37, 38].

3.10. Deficiency of vitamin C

Deficiency of vitamin C in humans causes a major disease called scurvy. The major signs of the disease occur primarily in mesenchymal tissues. It leads to impaired wound healing; oedema; haemorrhage (due to deficient formation of intercellular substance) in the skin, mucous membranes, internal organs, and muscles; and weakening of collagenous structures in bone, cartilage, teeth and connective tissues. Those who suffer from scurvy have swollen, bleeding gums with tooth loss. They also show lethargy, fatigue, rheumatic pains in the legs, muscular atrophy and skin lesions, massive sheet hematomas in the thighs, and ecchymoses and haemorrhages in many organs, including the intestines, sub-periosteal tissues and eyes. All these features are accompanied by psychological changes: hysteria, hypochondria and depression. In children, the syndrome is called Moeller-Barlow disease; it is seen in non-breastfed infants usually at about 6 months of age and is characterized by widening of bone-cartilage boundaries, stressed epiphyseal cartilage of the extremities, severe joint pain and frequently, anaemia and fever. Children having this disease present with a limp or inability to walk, tenderness of the lower limbs, bleeding of the gums and petechial haemorrhages.

4. Detection and sensing

Many analytical techniques are used for the determination of vitamin C in different matrices, such as titrimetric [39], fluorimetric [40], spectrophotometric [41], high-performance liquid chromatography [42], enzymatic [43], kinetic [44, 45] and electrochemical, etc.

4.1. High-performance liquid chromatography (HPLC)

High-performance liquid chromatography (HPLC) methods are preferred earlier because they are faster and more effective than spectrophotometric, titration or enzymatic methods, and they do not usually need derivatization [46]. In pharmaceutical and cosmetic industries, HPLC is used, which is considered as a sensitive and selective method.

Detection of vitamin C through HPLC has been done by many groups. Racz et al. in 1990 reported that HPLC used the HPLC method for the determination of ascorbic acid in fruits and vegetables. The method has been used to deep-frozen raspberry cream analysis together with three commonly used chemical methods of vitamin C analysis. The HPLC method has been compared with the chemical methods from several aspects, and the superiority of HPLC method has been concluded [47].

Snezana et al. reported the HPLC method for the determination of vitamin C in pharmaceutical samples in *Tropical Journal of Pharmaceutical Research* in 2011 [48]. The simplicity of this low-cost, rapid technique and its high specificity to ascorbic acid, even in the presence of a variety of excipients, demonstrate that this HPLC method would be particularly suitable for the determination of ascorbic acid (**Figure 9**). It can be used in pharmaceutical/veterinary formulations without prior sample preparation.

Another work carried out by Franco et al., which was to optimize and validate a new analytic strategy for the determination of vitamin C in strawberries by UV–HPLC [49].

For the accurate and reliable measurement of ascorbic acid and dehydroascorbic acid, HPLC can be used in combination with electrochemical or ultraviolet detection. Line and Hoffer have reported a method for the determination of vitamin C in plasma. It can be analysed by HPLC in connection with electrochemical or ultraviolet light detection. Electrochemical HPLC has advantages over UV-HPLC for plasma total vitamin C analysis [50].

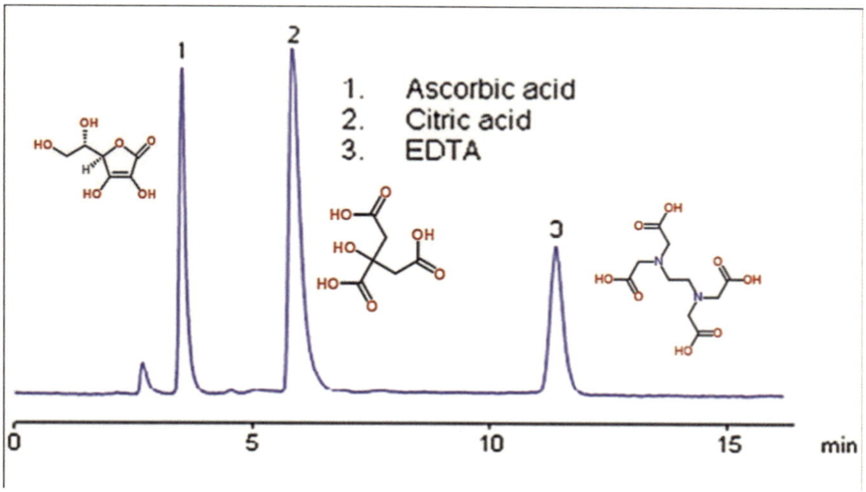

Figure 9. HPLC plot of ascorbic acid detection.

4.2. Spectrophotometric method

Among many analytical methods, spectrophotometric methods are very simple and low-cost. Several studies used the spectrophotometric method for the determination of ascorbic acid. Güçlü et al. [51] have proposed a spectrophotometric method based on ascorbic acid oxidation to dehydroascorbic acid, by using the Cu(II)-neocuproine complex, which is reduced to Cu(I)-bis(neocuproine), the absorbance of the latter being determined at 450 nm. Other optical methods for vitamin C estimation include spectrophotometrical determination of iodine reacted with ascorbic acid [52] and chemiluminescence [53].

A sensitive, simple and low-cost spectrophotometric method was introduced by Kobra and Somayye in 2015. The present method was successfully applied to determine the ascorbic acid in food and pharmaceutical samples. The samples were multivitamin tablet, effervescent tablet, vitamin C injection, natural orange juice, orange syrup powdered and commercial orange liquid. The method is based on the reaction of $AgNO_3$ with ascorbic acid in the presence of polyvinyl pyrrolidone (PVP) and slightly basic medium to prepare silver nanoparticles [54].

Kapur et al. in 2015 reported a method that is based on the oxidation of ascorbic acid to dehydroascorbic acid by bromine water in the presence of acetic acid. In this method, the total ascorbic acid (ascorbic acid + dehydroascorbic acid) has been determined in 21 different samples of fruits and vegetables by the spectrophotometric method [55]. Mohammed and Hazim in 2016 reported a UV-spectrophotometric method for the determination of ascorbic acid in fruits and vegetables from hill region with 2,4-dinitrophenylhydrazine [56].

A new sensitive colorimetric method for the determination of ascorbic acid tablet in aqueous solution was reported by Ahmed and Mohamed in 2013. The method is based on the formation of coloured azo dye by diazotization of 2,4-dichloroaniline, followed by azo-coupling reaction between the resulting product and ascorbic acid [57].

4.3. Electrochemical sensing

Electrochemical sensing methods are more widely applied as they are simple, sensitive and moderate methods. Ascorbic acid (AA), dopamine and catecholamine are important neurotransmitters in the human body. Hence, the low-level detection of ascorbic acid is very important. All the methods mentioned above involve complicated pretreatment techniques and expensive instruments. Hence, researchers have developed a more simple, sensitive and accurate electrochemical method for the detection of ascorbic acid. Several studies have reported the electrochemical sensing methods for detection of ascorbic acid. Attempts to simplify such methods have resulted in the development of new methods such as the nickel hexacyanoferrate film-modified aluminium electrode [58] and graphite epoxy electrode [59]. These methods use specific and modified working electrode systems that are complicated [60]. These electrochemical methods are widely used to monitor AA metabolites *in vivo* or *in vitro* [61, 62]. For example, batch injection amperometric methods can achieve a detection limit as low as 450 mg/L [63]. The metallic nickel electrode (2610–6 M), chemiluminometric flow method (51 M) [64], carbon paste mixed electrode (1.08610–5 M) [65], polyviologen-modified technique (0.381 M) [66] and NiGCNF plating-modified electrode (2610–6 M) [67] achieve very sensitive detection ranges.

Suw et al. in 2004 developed a stripping voltammetric technique that manifests faster response, more cost-effective and sensitive preconcentration techniques, which was used along with a simpler glassy carbon electrode [68]. In this method, low ascorbic acid concentrations were detected by the square wave stripping voltammetry and a glassy carbon electrode. The lower detection limit reported as 0.301 g/L (S/N = 3). This method can be used to detect biological materials, pharmaceuticals, food and drugs.

Many reports are there for the simultaneous electrochemical detection of ascorbic acid in the presence of dopamine and uric acid on the glassy carbon electrode. From our own group, we have synthesized electrically conducting poly (3,4-ethylenedioxythiophene) nanospindles (PEDOTSs) and used for the electrochemical sensing of ascorbic acid [69] (**Figure 10**).

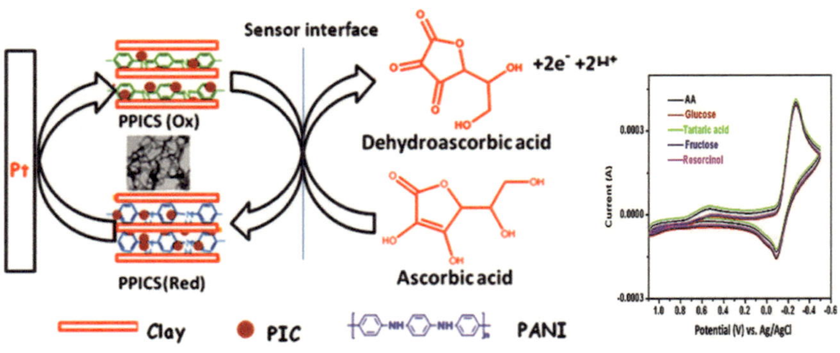

Figure 10. Cyclic voltammogram of electrochemical sensing of ascorbic acid using a polyaniline-modified platinum electrode [70].

In another study, our group developed a low-cost electrochemical sensor based on a platinum electrode for ascorbic acid conductive polyaniline-based composite. This unique low-cost and user-friendly sensor was validated for the nanomolar detection of AA. The lower detection limit for AA was observed at 0.1 nm. The aqueous PPICS/Pt electrode could serve as a prospective low-cost and efficient electrochemical sensor and provide promising and outstanding contributions to the food, beverage and medical industries [70].

Electrochemical methods have attracted much attention from clinical diagnostic perspectives because of their easy operation, low cost, rapid response, high sensitivity and good selectivity.

5. Conclusion

Vitamin C plays a pivotal role in body-building process and in disease prevention. The various functions of vitamin C, including the antioxidant activity, formation of protein, tendons, ligaments and blood vessels, for healing wounds and form scar tissue, for repairing and maintaining cartilage, bone, and teeth, and aiding in the absorption of iron, were discussed. This chapter will definitely benefit the students, researchers and technologists globally.

Author details

Sudha J. Devaki* and Reshma Lali Raveendran

*Address all correspondence to: sudhajd2001@yahoo.co.in

Chemical Sciences and Technology Division, CSIR-National Institute for Interdisciplinary Science and Technology (CSIR-NIIST), Thiruvananthapuram, India

References

[1] Food and Nutrition Board. Dietary Reference Intakes for Vitamin C, Vitamin E, Selenium, and Carotenoids. Washington, DC: National Academy Press; 2000

[2] Nutrient Data Laboratory. USDA National Nutrient Database for Standard Reference, Release 24 [Internet]. Available from http://www.ars.usda.gov/ba/bhnrc/nd

[3] Committee on Medical Aspects of Food Policy; 1991

[4] Food and Nutrition Board: Institute of Medicine; 2000

[5] Food and Agricultural Organization, World Health Organization; 2002

[6] Khan A, Rashid A, Younas R, Chong R. A chemical reduction approach to the synthesis of copper nanoparticles. International Nano Letters. 2016;**6**:21-26.

[7] Sun K, Qiu J, Liu J, Miao Y. Preparation and characterization of gold nanoparticles using ascorbic acid as reducing agent in reverse micelles. J Mater Sci. 2009;**44**:754

[8] D'souza SL, Pati RK, Kailasa SK. Ascorbic acid functionalized gold nanoparticles as a probe for colorimetric and visual read-out determination of dichlorvos in environmental samples. Analytical Methods. 2014;**6**:9007-9014

[9] Cameron E, Campbell A. The orthomolecular treatment of cancer. II. Clinical trial of high-dose ascorbic acid supplements in advanced human cancer. Chem Biol Interact. 1974;**9**:285-315

[10] Cameron E, Pauling L. Supplemental ascorbate in the supportive treatment of cancer: Prolongation of survival times in terminal human cancer. Proc Natl Acad Sci U S A. 1976; **73**:3685-9

[11] Cameron E, Pauling L. Supplemental ascorbate in the supportive treatment of cancer: reevaluation of prolongation of survival times in terminal human cancer. Proc Natl Acad Sci U S A. 1978;**75**:4538-42

[12] Chen Q, Espey MG, Sun AY, Pooput C, Kirk KL, Krishna MC. Pharmacologic doses of ascorbate act as a prooxidant and decrease growth of aggressive tumor xenografts in mice. Proceedings of the National Academy of Sciences of the United States of America. 2008;**105**:11105-11109

[13] Chen Q, Espey MG, Krishna MC, Mitchell JB, Corpe CP, Buettner GR. Pharmacologic ascorbic acid concentrations selectively kill cancer cells: Action as a pro-drug to deliver hydrogen peroxide to tissues. Proceedings of the National Academy of Sciences of the United States of America. 2005;**102**:13604-13609

[14] Chen Q, Espey MG, Sun AY, Lee JH, Krishna MC, Shacter E. Ascorbate in pharmacologic concentrations selectively generates ascorbate radical and hydrogen peroxide in extracellular fluid in vivo. Proceedings of the National Academy of Sciences of the United States of America. 2007;**104**:8749-8754

[15] Honarbakhsh S, Schachter M. Vitamins and cardiovascular disease. British Journal of Nutrition. 2008;**101**:1113-1131

[16] Jacob RA, Sotoudeh G. Vitamin C function and status in chronic disease. Nutrition in Clinical Care. 2002;**5**:66-74

[17] Willcox BJ, Curb JD, Rodriguez BL. Antioxidants in cardiovascular health and disease: Key lessons from epidemiologic studies. American Journal of Cardiology. 2008;**101**:75-86D

[18] Pauling L. The significance of the evidence about ascorbic acid and the common cold. Proceedings of the National Academy of Sciences of the United States of America. 1971;**68**:2678-2681

[19] Douglas RM, Hemilä H, Chalker E, Treacy B. Vitamin C for preventing and treating the common cold. Cochrane Database Systematic Review. 2007;3:CD000980

[20] Johnston CS. The antihistamine action of ascorbic acid. Subcellular Biochemistry. 1996;**25**:189-213

[21] Evans JR, Lawrenson JG. Antioxidant vitamin and mineral supplements for slowing the progression of age-related macular degeneration. Cochrane Database Systematic Review. 2012;11:CD000254

[22] Age-Related Eye Disease Study Research Group. A randomized, placebo-controlled, clinical trial of high-dose supplementation with vitamins C and E, beta carotene, and zinc for age-related macular degeneration and vision loss: AREDS report no. 8. Archives of Ophthalmology. 2001;**119**:1417-1436

[23] The Age-Related Eye Disease Study 2 (AREDS2) Research Group. Lutein + zeaxanthin and omega-3 fatty acids for age-related macular degeneration: The Age-Related Eye Disease Study 2 (AREDS2) randomized clinical trial. Journal of American Medical Association. 2013;**309**:2005-2015

[24] Carr AC, Frei B. Toward a new recommended dietary allowance for vitamin C based on antioxidant and health effects in humans. American Journal of Clinical Nutrition. 1999;**69**:1086-1107

[25] Rautiainen S, Lindblad BE, Morgenstern R, Wolk A. Vitamin C supplements and the risk of age-related cataract: A population-based prospective cohort study in women. American Journal of Clinical Nutrition. 2010;**91**(2):487-493

[26] Darr D, Combs S, Dunston S, Manning T, Pinnell S. Topical vitamin C protects porcine skin from ultraviolet radiation-induced damage. British Journal of Dermatology. 1992;**127**:247-253

[27] Steiling H, Longet K, Moodycliffe A. Sodium-dependent vitamin C transporter isoforms in skin: Distribution, kinetics, and effect of UVB-induced oxidative stress. Radical Biology and Medicine. 2007;**43**:752-762

[28] Kang JS, Kim HN, Jung da J. Regulation of UVB-induced IL-8 and MCP-1 production in skin keratinocytes by increasing vitamin C uptake via the redistribution of SVCT-1 from the cytosol to the membrane. Journal of Investigative Dermatology. 2007;**127**:698-706

[29] Tebbe B, Wu S, Geilen CC, Eberle J, Kodelja V, Orfanos CE. L-ascorbic acid inhibits UVAinduced lipid peroxidation and secretion of IL-1alpha and IL-6 in cultured human keratinocytes in vitro. Journal of Investigative Dermatology. 1997;**108**:302-306

[30] Eberlein-Konig B, Placzek M, Przybilla B. Protective effect against sunburn of combined systemic ascorbic acid (vitamin C) and d-alpha-tocopherol (vitamin E). Journal of the American Academy of Dermatology. 1998;**38**:45-48

[31] Bissett DL, Chatterjee R, Hannon DP. Photoprotective effect of superoxide-scavenging antioxidants against ultraviolet radiation-induced chronic skin damage in the hairless mouse. Photodermatology, Photoimmunology & Photomedicine. 1990;**7**:56-62

[32] Lin FH, Lin JY, Gupta RD. Ferulic acid stabilizes a solution of vitamins C and E and doubles its photoprotection of skin. Journal of Investigative Dermatology. 2005;**125**:826-832

[33] Lin JY, Selim MA, Shea CR. UV photoprotection by combination topical antioxidants vitamin C and vitamin E. Journal of the American Academy of Dermatology. 2003;**48**:866-874

[34] Savini I, Catani MV, Rossi A, Duranti G, Melino G, Avigliano L. Characterization of keratinocyte differentiation induced by ascorbic acid: Protein kinase C involvement and vitamin C homeostasis. Journal of Investigative Dermatology. 2002;**118**:372-379

[35] Boyce ST, Supp AP, Swope VB, Warden GD. Vitamin C regulates keratinocyte viability, epidermal barrier, and basement membrane in vitro, and reduces wound contraction after grafting of cultured skin substitutes. Journal of Investigative Dermatology. 2002;**118**:565-572

[36] Ponec M, Weerheim A, Kempenaar J. The formation of competent barrier lipids in reconstructed human epidermis requires the presence of vitamin C. Journal of Investigative Dermatology. 1997;**109**:348-355

[37] Silverstein RJ, Landsman AS. The effects of a moderate and high dose of vitamin C on wound healing in a controlled guinea pig model. Journal of Foot and Ankle Surgery. 1999;**38**:333-338

[38] Vaxman F, Olender S, Lambert A. Effect of pantothenic acid and ascorbic acid supplementation on human skin wound healing process. A double-blind, prospective and randomized trial. Journal of Surgical Research. 1995;**27**:158-166

[39] Verma KK, Jain A, Sahasrabuddhey B, Gupta K, Mishra S. Solid-phase extraction cleanup for determining ascorbic acid and dehydroascorbic acid by titration with 2,6-dichlorophenolindophenol. Journal of AOAC International. 1996;**79**:1236

[40] Xia W, Yuxia D, Changxia S, Jinghe Y, Yuebo W, Shuna S. Fluorimetric determination of ascorbic acid with o-phenylenediamine. Talanta. 2003;**59**:95-99

[41] Rahman Khan MM, Rahman MM, Islam MS, Begum SA. A simple UV-spectrophotometric method for the determination of vitamin C content in various fruits and vegetables at Sylhet area in Bangladesh. Journal of Biological Sciences. 2006;**6**:388-392

[42] Nyyssönen K, Salonen JT, Parviaine MT. Ascorbic acid. In: De Leenheer AP, Lambert WE, Van Bocxlaer JF, editors. Modern Chromatographic Analysis of Vitamins. 3rd ed. Marcel Dekker: New York; 2000

[43] Casella L, Gullotti M, Marchesini M, Petrarulo M. Rapid enzymatic method for vitamin C assay in fruits and vegetables using peroxidase. Journal of Food Science. 2006;**54**:374

[44] Ensafi AA, Rezaei B, Movahedinia H. Kinetic–spectrophotometric determination of ascorbic acid by inhibition of the hydrochloric acid–bromate reaction. Spectrochimica Acta Part A: Molecular and Biomolecular Spectroscopy. 2002;**58**:2589-2594

[45] Safavi A, Fotouhi L. Kinetic spectrophotometric determination of ascorbic acid by reduction of toluidine blue. Talanta. 1994;**41**:1225-1228

[46] Nováková L, Solich P, Solichová D. HPLC methods for simultaneous determination of ascorbic and dehydroascorbic acids. Trends in Analytical Chemistry. 2008;**27**:942

[47] Racz E, Parlagh-Huszar K, Kecskes T. HPLC method for determination of ascorbic acid in fruits and vegetables. Periodica Polytechnica Chemical Engineering. 1991;**351**-2:23-30

[48] Snezana SM, Danijela AK, Danijela CN, Milan NM. Rapid and reliable HPLC method for the determination of vitamin C in pharmaceutical samples. Tropical Journal of Pharmaceutical Research. 2011;**10**(1):105-111

[49] Franco VV, María EP, María SC, Daniel RG, del H. Bernardi CM. Optimization and validation of a UV–HPLC method for vitamin C determination in strawberries (Fragaria ananassa Duch.), using experimental designs. Food Analytical Methods. 2012;**5**:1097-1104

[50] Line R, Hoffer LJ. A simple method for plasma total vitamin C analysis suitable for routine clinical laboratory use. Nutrition Journal. 2016;**15**:40

[51] Güçlü K, Sözgen K, Tütem E, Özyürek M, Apak R. Spectrophotometric determination of ascorbic acid using copper(II)-neocuproine reagent in beverages and pharmaceuticals. Talanta. 2005;**65**(5):1226-1232

[52] Danet AF, David V, Oancea M. Dispozitiv de analiza in flux cu injectarehidrodinamica. Determinareavitaminei C, Revista de Chimie. 1994;**45**(11):1000-1006

[53] Danet BM, Aboul-Enein HY. Flow injection system with chemiluminometric detection for enzymatic determination of ascorbic acid. Luminescence. 2000;**155**:305-309

[54] Kobra Z, Somayye M. Sensitive spectrophotometric determination of ascorbic acid in drugs and foods using surface plasmon resonance band of silver nanoparticles. Cogent Chemistry. 2015;**1**:1109172

[55] Kapur A, Hasković A, Čopra-Janićijević A, Klepo L, Topčagić A, Tahirović I, Sofić E. Spectrophotometric analysis of total ascorbic acid contetnt in various fruits and vegetables. Bulletin of the Chemists and Technologists of Bosnia and Herzegovina. 2012;**38**:39-42

[56] Mohammed IHM, Hazim YA. Determination of Vitamin C (ascorbic acid) Contents in various fruit and vegetable by UV-spectrophotometry and titration methods. Journal of Chemical and Pharmaceutical Sciences. 2016;**9**:2972

[57] Ahmed Z, Mohamed H. Spectrophotometric determination of ascorbic acid in aqueous solutions and in pharmaceuticals formulations. Journal of Al-Nahrain University. 2013;**16**:65-71

[58] Cooper JA, Woodhouse KE, Chippindale AM, Compton RG. Photoelectrochemical determination of ascorbic acid using methylene blue immobilized in α-zirconium phosphate. Electroanalysis. 1999;**11**:1259-1265

[59] Morais H, Rodrigues P, Ramos C, Forgcs E, Cserhti T, Oliveira J. Effect of ascorbic acid on the stability of β-carotene and capsanthin in paprika (*Capsicum annuum*) powder. Nahrung/Food. 2002;**46**:308-310

[60] Julio CBF, Lauro TK, Graciliano DON. Potentiometric sensor for L-ascorbic acid based on EVA membrane doped with copper (II). Electroanalysis. 1999;**11**:475-480

[61] Song Z, Qun X, Wen Z, Litong J, Ji YJ. In vivo monitoring of the monoamine neurotransmitters in rat brain using microdialysis sampling with liquid chromatography electrochemical detection. Analytica Chimica Acta. 2001;**427**:45-53

[62] Joseph W, Alain W. Zeolite-modified carbon paste electrode for selective monitoring of dopamine. Journal of Electroanalytical. Chemistry. 1996;**407**:183-187

[63] Adriana DD, Jairo JP, Ivano GRG. A batch injection analysis system for ascorbic acid determination using amperometric detection on a sessile mercury drop electrode. Electroanalysis. 1999;**11**:1124-1129

[64] Andrei DF, Mihaela B, Hassan YAE. Flow injection system with chemiluminometric detection for enzymatic determination of ascorbic acid. Lumin. 2000;**15**:305-309

[65] Jahan BR, Reza O, Abolfazl K. Carbon paste electrode spiked with ferrocene carboxylic acid and its application to the electrocatalytic determination of ascorbic acid. Journal of. Electroanalytical Chemistry. 2001;**515**:45-51

[66] Jyh MZ, Dong MT, Hsueh HY. Direct determination of vitamin C in fruit juices using a polyviologen-modified electrode. Electroanalysis. 2002;**14**:1597-1600

[67] Pournaghi AMH, Razmi NH, Hafezi B. Amperometric determination of ascorbic acid in real samples using an aluminum electrode, modified with Nickel Hexacyanoferrate Films by simple electroless dipping method. Electroanalysis. 2002;**14**:206-219

[68] Suw YL, Jung IC, Young SJ, Woon WJ, Hye JL, Seong HL. Electrochemical detection of ascorbic acid (vitamin C) using a glassy carbon electrode. Nahrung/Food. 2004;**3**:201-204

[69] Sudha JD, Neethu KS, Renjith S, Hans-Juergen PA, Pich A. Water dispersible electrically conductive poly (3,4-ethylenedioxythiophene) nanospindles by liquid crystalline template assisted polymerization. Journal of Materials Chemistry C. 2014;**2**:6991

[70] Reena VL, Sudha JD, Rohini KN, Neethu KS. Design of a nanostructured electromagnetic polyaniline–keggin iron–clay composite modified electrochemical sensor for the nanomolar detection of ascorbic acid. Journal of Applied Polymer Science. 2014;**131**:40936-4094

Section 2

Antioxidant Effect of Vitamin C

Chapter 2

Vitamin C: An Antioxidant Agent

Fadime Eryılmaz Pehlivan

Additional information is available at the end of the chapter

http://dx.doi.org/10.5772/intechopen.69660

Abstract

Vitamin C or ascorbic acid (AsA) is a naturally occurring organic compound with antioxidant properties, found in both animals and plants. It functions as a redox buffer which can reduce, and thereby neutralize, reactive oxygen species. It is a cofactor for enzymes involved in regulating photosynthesis, hormone biosynthesis, and regenerating other antioxidants; which also regulates cell division and growth, is involved in signal transduction, and has roles in several physiological processes, such as immune stimulation, synthesis of collagen, hormones, neurotransmitters, and iron absorption, has also roles in detoxifying the body of heavy metals. Severe deficiency of vitamin C causes scurvy, whereas limited vitamin C intake causes symptoms, such as increased susceptibility to infections, loosening of teeth, dryness of the mouth and eyes, loss of hair, dry itchy skin, fatigue, and insomnia. In contrast, vitamin C can also act as a prooxidant, especially in the presence of transition metals, such as iron and copper, starting different hazardous radical reactions. Vitamin C can both act as a strong, efficient, and cheap antioxidant agent and, at the same time, behave as a radical promoter. Further investigations are needed to illuminate the dual roles of vitamin C

Keywords: vitamin C, antioxidant, ascorbic acid metabolism, prooxidant

1. Introduction

Vitamin C (L-ascorbic acid) is a water-soluble micronutrient required for multiple biological functions. It is necessary for normal growth and development, and is an essential enzyme cofactor for several enzymes in the post-translational hydroxylation of collagen, biosynthesis of carnitine, conversion of the neurotransmitter dopamine to norepinephrine, peptide amidation, and in tyrosine metabolism. It is also an antioxidant that helps protection against infection and iron absorption. Some animal species have lost the capacity for L-ascorbate synthesis, for that reason, they are dependent upon diet to ensure adequate levels of vitamin C for metabolism and oxidative protection. The high L-ascorbate contents found in plants make them the primary source of vitamin C intake for humans [1–3]. Vitamin C is one of the potent reducing agents and scavenger of free radicals in biological systems, working as a scavenger of oxidizing free radicals and harmful oxygen-derived species, such as hydroxyl radical, hydrogen peroxide (H_2O_2), and singlet oxygen [3, 4]. Many uses for vitamin C have been proposed, such as its antiscorbutic action, but few have been found to be beneficial in scientific studies [1–4]. In particular, research in stomach cancer, and other cancers, asthma, diabetes, cataracts, or heart disease remains inconclusive. In plants, vitamin C has roles in processes such as growth, programmed cell death, pathogen responses, hormone responses, flowering, and senescence, as well as protection against environmental stresses [1–5]. Vitamin C also plays an important role in abiotic stress tolerance, and considerable interest has focused on it due to its ability to induce a protective effect on plants under stress. It has been supported that vitamin C induced increases in the resistance of plants on heavy metal stress [5]. The role of exogenously applied ascorbic acid (AsA) under heavy metal stress on the photosynthetic pigments, membrane permeability, and mineral uptake of plants is not still clear [5, 6].

2. Biosynthesis and molecular structure of vitamin C

In animals, biosynthesis of vitamin C is included in the glucuronic acid metabolic pathway, which is involved in the metabolism of sugars under normal and disease conditions, and in regulation of physiological functions (**Figure 1a**). Glucuronic acid metabolic pathway is also an important pathway for detoxification processes [3, 4]. While most animals can convert D-glucose into L-ascorbic acid, humans and other primates, guinea pigs, some fish and birds, and insects are unable to produce ascorbic acid endogenously. The major plant pathway is different from the animal L-ascorbate synthesis pathway that involves 10 enzymatic steps from D-glucose to L-ascorbate via the intermediate formation of GDP-D-mannose and L-galactose [3] (**Figure 1a**).

Vitamin C (L-ascorbic acid) is a dibasic acid with an enediol group built into a five-membered heterocyclic lactone ring (**Figure 1b**).

The chemical and physical properties of ascorbic acid are related to its structure [7]. The structure of dehydroascorbic acid, the first oxidation product of ascorbic acid, has been analyzed by X-ray crystallography to be a dimer (**Figure 1c**).

Electrochemical studies have indicated that ascorbic acid and dehydroascorbic acid form a reversible redox couple (**Figure 1d**).

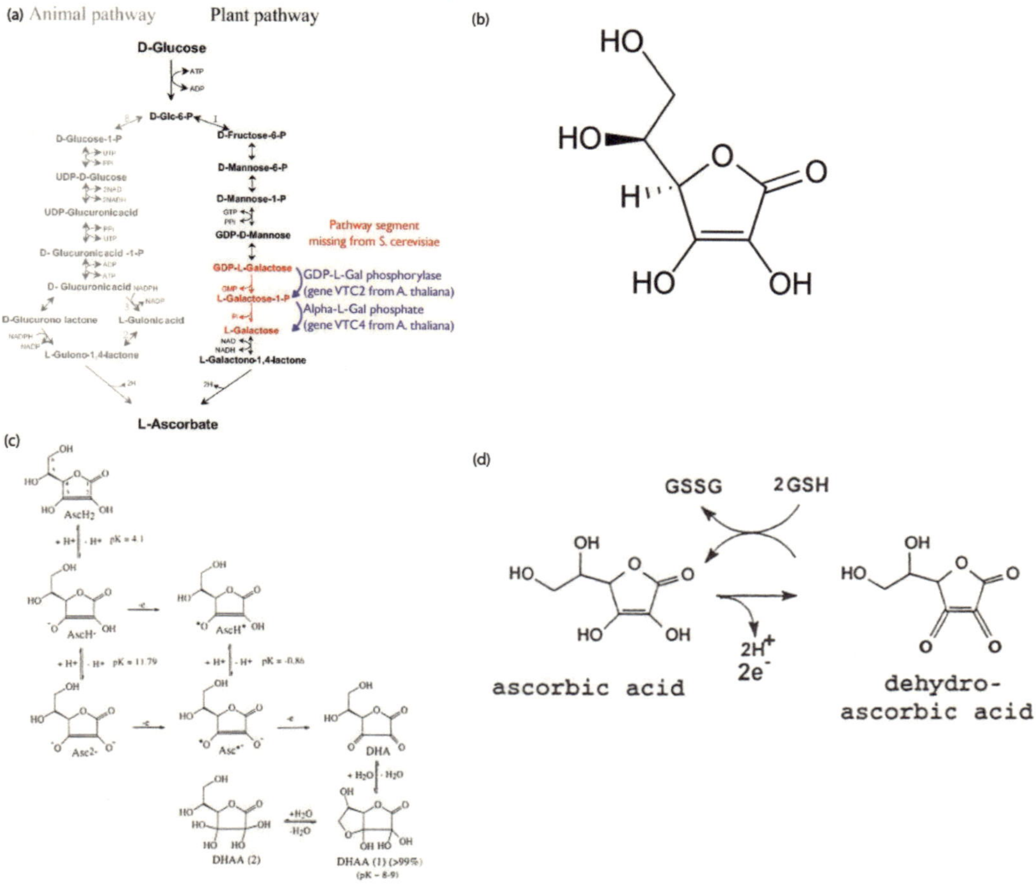

Figure 1. (a) Diagram for the vitamin C pathway [6]; (b) molecular structure of L-ascorbic acid (vitamin C); (c) the equilibrium and redox species in the ascorbic acid-dehydroascorbic acid system; (d) ascorbic acid and dehydroascorbic acid. The oxidized form, dehydroascorbic acid, can be reduced back to ascorbic acid by glutathione (GSH).

3. Redox metabolism and antioxidant properties of vitamin C

Free radicals and oxidants play a dual role as both toxic and beneficial compounds, in metabolic processes and in response to exogenous stimulations. They are produced either from normal metabolic activities or from environmental factors (pollution, cigarette smoke, and radiation). When an overload of free radicals cannot be scavenged, their accumulation in the body generates oxidative stress [3]. Oxidative stress occurs when free radical formation exceeds the ability of protection against them. This process leads in the development of chronic and degenerative illnesses such as cancer, autoimmune disorders, aging, cataract, rheumatoid arthritis, cardiovascular, and neurodegenerative diseases [8–10]. An antioxidant is a molecule that prevents the oxidation of other molecules. Oxidation process is a chemical reaction that produces free radicals, leading to chain reactions that may damage cells. The antioxidant effect of vitamin C has been well documented [8, 9, 11]. Vitamin C is a powerful antioxidant having ability to donate a hydrogen atom and form a relatively stable ascorbyl-free radical. Vitamin E, vitamin C, and β-carotene are known as antioxidant vitamins that are suggested to decrease oxidative damage and lowering the risk of certain chronic diseases. Diseases, such as cardiovascular disorders, are associated with inadequate concentrations of L-ascorbic acid, tocopherol, and β-carotene [8–10] in epidemiological studies. Vitamin C also enhances iron absorption by reducing Fe^{3+} to Fe^{2+} from non-heme iron sources [3, 23]. In the presence of redox-active ions (iron, copper), vitamin C acts as a prooxidant, contributing to the formation of hydroxyl radicals, that may lead to lipid, DNA, or protein oxidation [8–10]. There are different mechanisms to alleviate oxidative stress and repair damaged macromolecules. Enzymatic and nonenzymatic antioxidants have important roles in scavenging free radicals and reactive oxygen species (ROS). The antioxidant enzymes, catalase (CAT), superoxide dismutase (SOD), glutathione reductase (GR), glutathione peroxidase (GSHpx) and, in plants, ascorbate peroxidase (AA-px) and the nonenzymatic antioxidants, including glutathione (GSH) and ascorbate (ASC), have been shown to be significantly affected by oxidative stress [8, 9]. Antioxidant compounds can prevent the uncontrolled formation of free radicals or inhibit their reaction with biological sites; also, the destruction of most free radicals depends on the oxidation of endogenous antioxidants mainly by scavenging and reducing molecules [8, 9]. Vitamin C is thought to be an important water soluble antioxidant which is reported to neutralize ROS and reduce the oxidative stress [8, 10].

Vitamin C is a potent reducing agent and scavenger of free radicals in biological systems [11]. It is involved in the first line of antioxidant defense, protecting lipid membranes, and proteins from oxidative damage. As a water soluble molecule, vitamin C can work both inside and outside the cells, and can neutralize free radicals and prevent free radical damage. Vitamin C is an excellent source of electrons for free radicals that are seeking out an electron to regain their stability. Vitamin C can donate electrons to free radicals and quench their reactivity [8, 9].

Vitamin C has been shown to be an effective scavenger against oxygen and nitrogen oxide species, such as superoxide radical ion, hydrogen peroxide, the hydroxyl radical, and singlet oxygen. This property of vitamin C has vital processes in protection of cellular components from free radical-induced damage. In addition, vitamin C is effective in regenerating the antioxidant form of vitamin E by reducing tocopheroxyl radicals. This process protects membranes and other compartments of the cell from free radical-induced damage [8, 9] (**Figure 2**). Ascorbate peroxidase (APX) is an enzyme reducing H_2O_2 to water by using ascorbate as an electron donor. Monodehydroascorbate is an oxidized ascorbate that is regenerated by monodehydroascorbate reductase (MDAR). Monodehydroascorbate radical rapidly disproportionates into ascorbate and dehydroascorbate. Dehydroascorbate is reduced to ascorbate by dehydroascorbate reductase in the presence of GSH, yielding oxidized glutathione (GSSG). It is reduced by glutathione reductase (GR) using nicotinamide adenine dinucleotide phosphate hydrogen (NADPH) as an electron donor. Dehydroascorbate may be reduced nonenzymatically or catalyzed by proteins with dehydroascorbate reductase (DHAR) activity.

Glutathione-ascorbate cycle operates in the cytosol, mitochondria, plastids, and peroxisomes in plants [8, 9]. It is suggested that the glutathione-ascorbate cycle plays a key role for H_2O_2

http://dx.doi.org/10.5772/intechopen.69660

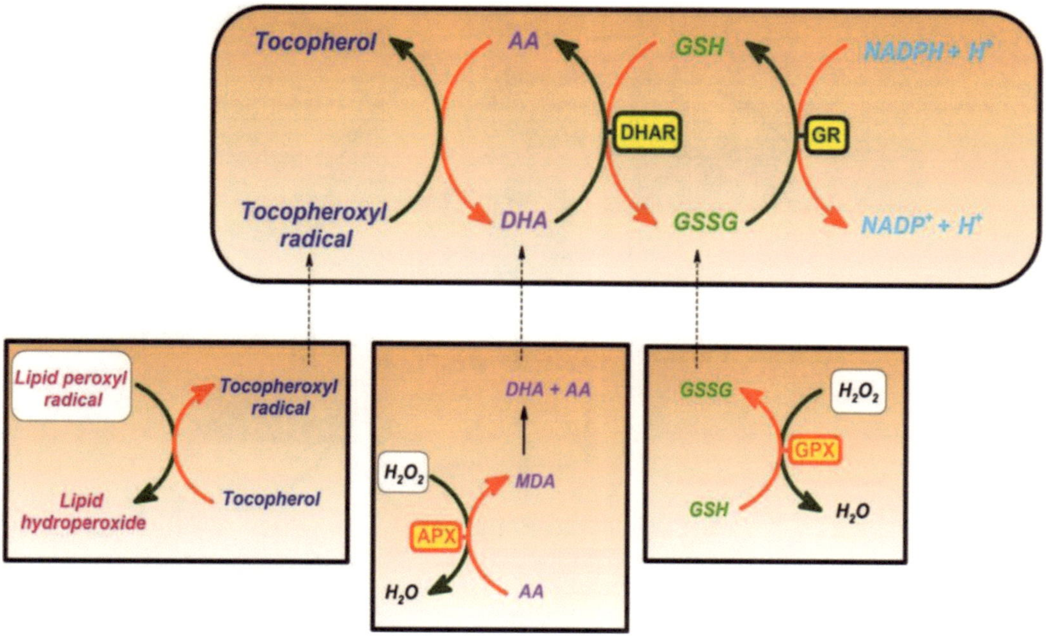

Figure 2. Ascorbate and redox cycling antioxidants. AA, ascorbate; DHA, dehydroascorbate; DHAR, semidehydroascorbate reductase; GSH, glutathione; GSSG, semi-glutathione reductase; GR, glutathione reductase; APX, ascorbate peroxidase; and GPX, glutathione peroxidase [12].

detoxification, because of the high concentrations of glutathione, ascorbate, and NADPH in plant cells. Other enzymes, such as ascorbate and glutathione peroxidases, which use thioredoxins or glutaredoxins as reducing substrates, also take roles in the removal of H_2O_2 in plants [8, 9] (**Figure 2**).

Vitamin C also forms the semidehydroascorbyl radical, a relatively long-lived radical, in regenerating vitamin E from its radical form, as well as in scavenging radicals. Plant and animal cells contain an NADH-dependent semidehydroascorbate reductase enzyme (EC 1.6.5.4), reducing the radical back to vitamin C by using NADH as a source of reducing agent (**Figure 2**). Both enzymatically and nonenzymatically, it can irreversibly decompose into diketogluconic acid or it can be converted to ascorbate in a glutathione-dependent reaction [3, 13, 14].

As being a reducing substance and an electron donor, during free radical scavenging, vitamin C donates high-energy electrons to neutralize free radicals, and it is oxidized to dehydroascorbic acid. Dehydroascorbic acid may be converted back into ascorbic acid for reuse or may be metabolized, further releasing more electrons. Although vitamin C is absorbed from the gut via a sodium-dependent vitamin C transporter, most cells transport vitamin C in an oxidized form (dehydroascorbic acid) via glucose transporter 1. Dehydroascorbic acid is reduced to generate ascorbic acid inside the cell, protecting mitochondria from free radical-induced oxidative damage (**Figures 2** and **3**). Highly reactive free radicals (e.g., RO^-, RO^{2-}, OH^-,, NO_2) are reduced by ascorbate, and the newly generated ascorbyl radical

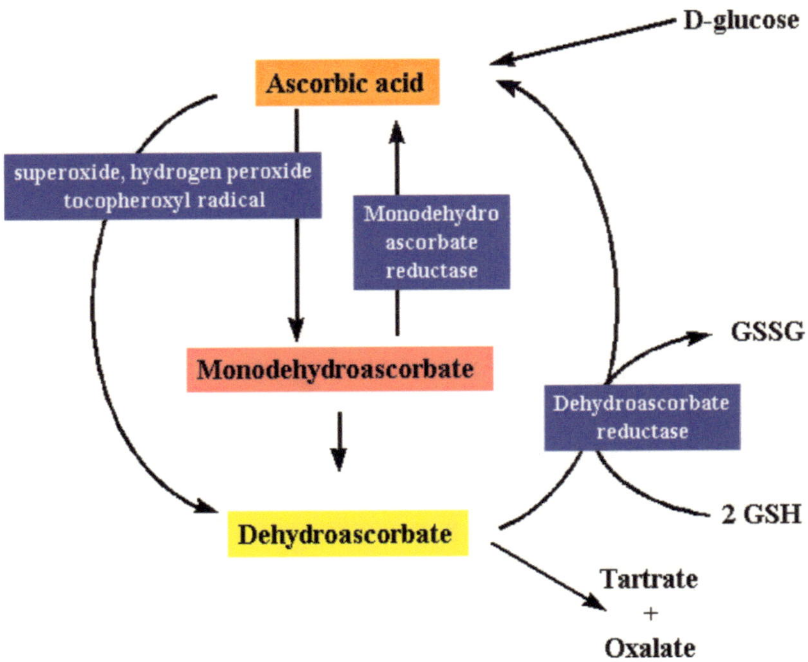

Figure 3. Synthesis and degradation of L-ascorbic acid in plant tissues [18].

is poorly reactive. Ascorbate can also scavenge nonradical reactive species, derived from peroxynitrite, such as hypochlorous acid, ozone, and nitrating agents. Vitamin C is a monosaccharide oxidation-reduction (redox) catalyst found in both animals and plants. The antioxidant effect of vitamin C is due to its ability to donate electrons from both the second and third carbon. During primate evolution, one of the enzymes needed to make ascorbic acid has been lost by mutation, humans must obtain it from the diet [15]; most animals can synthesize this vitamin in their bodies and do not require it in their diets [16]. Vitamin C is needed in the conversion of the procollagen to collagen by oxidizing proline residues to hydroxyproline. In other cells, it is maintained in its reduced form by reaction with glutathione [17]. As shown in **Figures 2** and **3**, ascorbic acid is a redox catalyst which can reduce, and thereby neutralize, ROS such as hydrogen peroxide (H_2O_2) (**Figures 2** and **3**).

Ascorbic acid has direct antioxidant effects, and also it is a substrate for the redox enzyme ascorbate peroxidase, that is particularly important in stress resistance in plants. Ascorbic acid is present at high levels in all parts of plants, especially in chloroplasts that reach concentrations of 20 mM there [19]. Dehydroascorbate (DHA) and ascorbate free radical (AFR), as an intermediate, the ascorbate free radical (AFR), that are reversible, one-electron oxidations are generated from ascorbate (**Figure 4**). According to the generally assumed model of enzymatic removal of ROS, SOD catalyzes superoxide anion to H_2O_2 and oxygen; then H_2O_2 is reduced into water and molecular oxygen by CAT. CAT turnover number is very high, but its affinity for H_2O_2 is relatively low, and consequently a certain amount of H_2O_2 remains in the cell.

H_2O_2 can react with superoxide anion formed in oxidative metabolism generating the highly reactive hydroxyl radical. GSH peroxidases (GSH-px) and AA peroxidases (AA-px) are capable

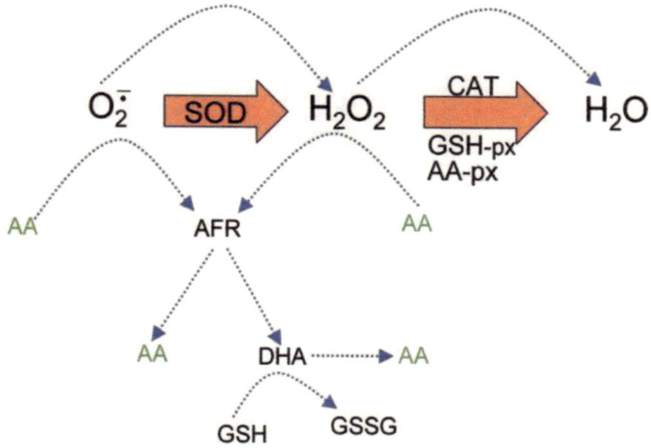

Figure 4. The role of AA in the detoxification of ROS. Blue dotted lines indicate nonenzymatic reactions.

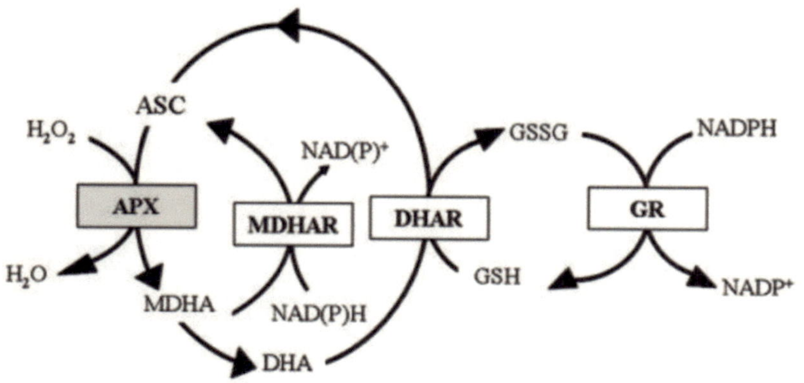

Figure 5. Foyer-Halliwell-Asada cycle [20].

of scavenging H_2O_2 due to their high affinity for H_2O_2. The cooperativity of SOD, CAT, and peroxidases ensures low amounts of superoxide anion and H_2O_2 and limiting the risk of hydroxyl radical formation (**Figure 5**).

4. Role of vitamin C in lipid peroxidation

The chemical and biological properties of L-ascorbic acid suggest that it can act as an antioxidant *in vivo* [21]. Vitamin C is a primary antioxidant in that it directly neutralizes radical species. It is not very reactive with prevalent cellular oxidants such as hydrogen peroxide and probably reacts mostly with hydrogen peroxide breakdown products [22]. Vitamin C has the ability to protect against lipid peroxidation by acting as a scavenger of ROS and by one-electron reduction of lipid hydroperoxyl radicals via the vitamin E redox cycle [23] (**Figure 6**).

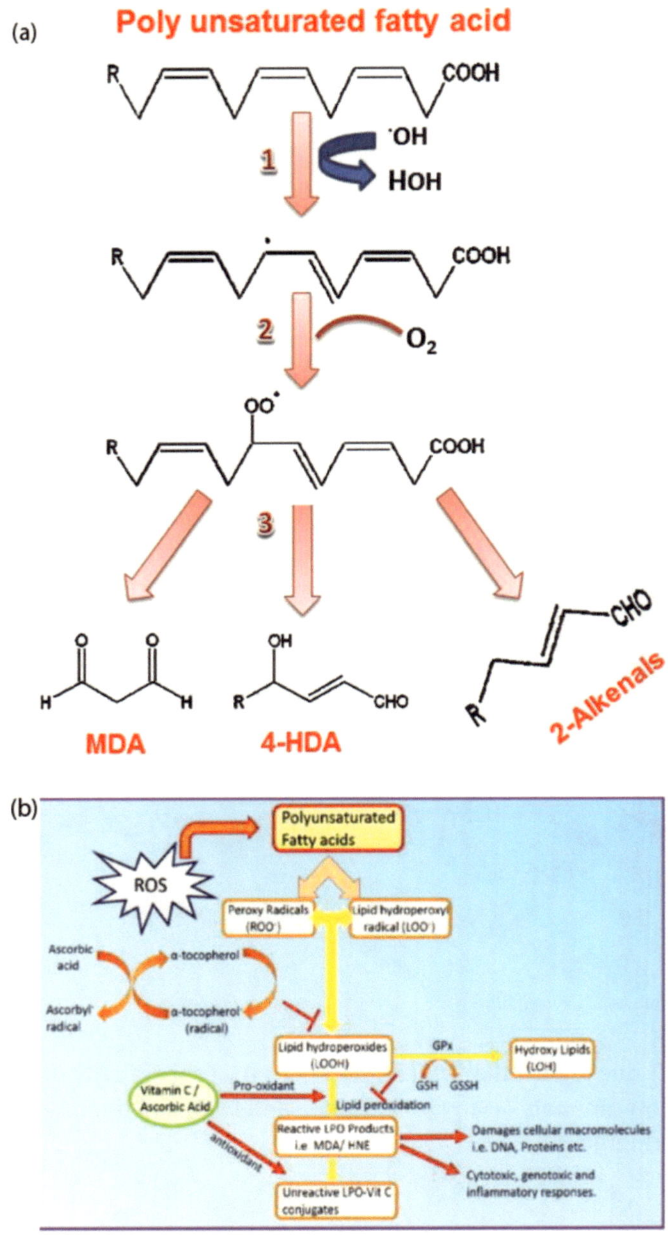

Figure 6. Schematic presentation of ROS-mediated lipid peroxidation chain reaction. Vitamin C serves dual role of a prooxidant and an antioxidant [23].

$LPO^{\bullet}$ radical can damage the macromolecules (DNA, RNA, and proteins) and can initiate cytotoxic, genotoxic, and inflammatory reactions. Vitamin C converts lipid peroxidation products into unreactive vitamin C-LPO products. This helps to prevent the interaction of macromolecules (DNA, RNA, and proteins) with $LPO^{\bullet}$ radicals. Vitamin C plays role in the regeneration of vitamin E; it donates electron to tocopheryl radical (vitamin-E–O•) and reduces it to tocopherol.

5. Prooxidant effect of vitamin C

The reducing agents, antioxidants, can also act as prooxidants. Vitamin C is also known to act as a prooxidant *in vitro*. Antioxidant vitamin C reduces oxidizing substances, such as hydrogen peroxide; however, it also reduces metal ions that generate free radicals through the Fenton reaction [25].

$$2Fe^{3+} + \text{ascorbate} \rightarrow 2Fe^{2+} + \text{dehydroascorbate} \tag{1}$$

$$2Fe^{2+} + 2H_2O_2 \rightarrow 2Fe^{3+} + 2OH\bullet + 2OH^- \tag{2}$$

Vitamin C is a reducing agent and antioxidant, and reacts with reactive oxygen species, such as the hydroxyl radical. Oxidative modifications of lipids, proteins, and DNA are induced by mixtures of ascorbic acid and copper or iron for decades [24, 25], contributing to oxidative damage formation by reducing ferric Fe^{3+} to ferrous Fe^{2+} ions (and Cu^{2+} to Cu^{+}), which in turn can reduce hydrogen peroxide (H_2O_2) to hydroxyl radicals. Therefore, vitamin C-mediated Fenton reactions should be controlled in the human body due to efficient iron sequestration by metal-binding proteins such as ferritin and transferrin. It has been suggested that the prooxidant effect may not be relevant *in vivo* [26]. Such radicals damaged macromolecules interacting with nucleic acids, proteins, and lipids, and initiating chain reactions. Ascorbate can terminate these chain radical reactions by electron transfer. The net reaction is

$$\begin{array}{l} RO\bullet + C \\ 6H \\ 7O^- \\ 6 \rightarrow RO^- + C_6H_7O\bullet \\ 6 \rightarrow ROH + C_6H_6O_6 \end{array} \tag{3}$$

The oxidized forms of ascorbate are relatively unreactive and do not cause cellular damage.

In the presence of free metal ions, excess ascorbate promotes and initiates free radical reactions. This is a potentially dangerous prooxidative compound; thus, vitamin C supplements are not recommended in people with high iron levels [3, 27]. It is provided a mechanism which, vitamin C induces the decomposition of lipid hydroperoxides to genotoxic bifunctional electrophiles *in vitro* without the need for free transition metal ions [3, 27].

6. Vitamin C in human disease

As an electron donor, vitamin C could be involved in several disease processes. Vitamin C is present in almost all foods of plant origin. The minimal vitamin C requirement for humans is defined as 40–60 mg/day to combat dietary deficiency [28]. However, vitamin C status decreases with both age and smoking, and is associated with chronic diseases such as rheumatoid arthritis and cancer [28]. Vitamin C might be consumed by preventing free radical-induced damage of DNA, which is thought to be an initiating step in cancer formation. The possible use of vitamin C in cancer therapy and prevention has been an area of great interest. Vitamin C supplements, which are able to prevent the formation and/or promote the repair of pre-mutagenic oxidative DNA lesions, are suggested to be of use in cancer prevention. Recently, a report showed that daily supplementation with vitamin C at high doses increased the survival time of terminal cancer patients, suggesting that vitamin C can have important anticancer properties. Indeed, vitamin C is proved to kill or inhibit the growth of many tumor cell lines [28]. Regarding cancer prevention, several epidemiological studies have linked the consumption of a diet rich in fruit and vegetables with lower incidence of many types of cancer [3, 29–34]. The hypothesis supports oxidative processes regulate ascorbate catabolism in humans, although direct proof is currently lacking. Ascorbate plays an important role as a first defense against oxidative stress. Smokers present one example of the relationship between oxidants and ascorbate since they expose themselves to oxidants via inhaled smoke. These oxidants have been demonstrated to induce lipid peroxidation *in vitro*, which is prevented by the presence of ascorbate. It is demonstrated that the turnover of 14C-labeled ascorbate has been 40% greater in smokers compared with nonsmokers. Ascorbate requirements of smokers based on intake and serum concentrations show that smokers may require as much as three times the dose of ascorbate than nonsmokers to avoid risk of efficiency [30–32, 34].

7. Conclusion

Vitamin C, as an antioxidant agent, has been the object of several investigations. However, vitamin C can also act as a prooxidant, especially in the presence of transition metals such as iron and copper, starting different dangerous radical reactions. As the experimental findings that have not been obtained by studying *in vitro* systems have been confirmed by in vivo investigations, further work is strongly needed. Consequently, ascorbic acid can act as a strong, efficient, and cheap antioxidant agent, whereas at the same time it can behave as a radical promoter and can produce dangerous species in living systems. Further investigations are needed to clear the data in the literature completely.

Author details

Fadime Eryılmaz Pehlivan

Address all correspondence to: eryilmazfadime@gmail.com

Istanbul University, Faculty of Science, Department of Biology, Botany Section, Süleymaniye, Fatih, Istanbul, Turkey

References

[1] Tolbert BM, Downing M, Carlson RW. Chemistry and metabolism of ascorbic acid and ascorbate sulfate. The Annals of the New York Academy of Science. 1975;**258**:48-69

[2] Packer L, Fuchs J. Vitamin C in Health and Disease. New York: Marcel Dekker Inc.; 1997

[3] Hacışevki A. An overview of ascorbic acid biochemistry. Journal of Faculty of Pharmacy. 2009;**38**(3):233-255

[4] Arrigoni O, De Tullio MC. Ascorbic acid: Much more than just an antioxidant. Biochimica et Biophysica Acta, 2002;**1569**:1-9

[5] Eryılmaz Pehlivan F. Effects of exogenously applied ascorbic acid on red cabbage cotyledons subjected to copper excess. Fresenius Environmental Bulletin. 2014;**23**:15-19

[6] Sauer M, Branduardi P, Valli M, Porro D. Production of L-ascorbic acid by metabolically engineered *Saccharomyces cerevisiae* and *Zygosaccharomyces bailii*. Applied and Environmental Microbiology. 2004;**70**(10):6086-6091

[7] Aguirre R, May JM. Inflammation in the vascular bed: Importance of vitamin C. Pharmacology & Therapeutics. 2008;**119**:96-103

[8] Rouhier N Lemaire SD, Jacquot JP. The role of glutathione in photosynthetic organisms: Emerging functions for glutaredoxins and glutathionylation. Annual Review of Plant Biology. 2008;**59**:143-166. DOI:10.1146/annurev.arplant.59.032607.092811. PMID 18444899.

[9] Bindhumol V, Chitra KC, Mathur PP. A induces reactive oxygen species generation in the liver of male rats. Toxicology. 2003;**188**:117-124

[10] Verma RS, Mhta A, Srivastava N. In vivo chlorpyrifos induced oxidative stress: Attenuation by antioxidant vitamins. Pesticide Biochemistry and Physiology. 2007;**88**:191-196

[11] Duarte TL, Lunec J. Review: When is an antioxidant not an antioxidant? A review of novel actions and reactions of vitamin C. Free Radical Research. 2005;**39**(7):671-686

[12] Szarka A, Tomassovics B, Bánheghyi G. The ascorbate-glutathione-α-tocopherol triad in abiotic stress response. International Journal of Molecular Sciences. 2012;**13**:4458-4483. DOI: 10.3390/ijms13044458

[13] Doba TG, Burton W, Ingold KU. Antioxidant and co-antioxidant activity of vitamin C: The effect of vitamin C, either alone or in the presence of vitamin E or a water-soluble vitamin E analogue, upon the peroxidation of aqueous multilamellar phospholipid liposomes. Biochimica et Biophysica Acta. 1985;**835**:298-303

[14] Winkler BS, Orselli SM, Rex TS. The redox couple between glutathione and ascorbic acid: A chemical and physiological perspective. Free Radical Biology and Medicine. 1994;**17**:333-349

[15] Khaw KT, Bingham S, Welch A, Luben R, Wareham N, Oakes S, Day N. Relation between plasma ascorbic acid and mortality in men and women in Epic-Norfolk prospective study: A prospective population study. Lancet. 2001;**357**:657-663

[16] May JM. How does ascorbic acid prevent endothelial dysfunction? Free Radical Biology and Medicine. 2000;**28**:1421-1429

[17] Huang A, Vita JA, Venema RC, Keaney JF. Ascorbic acid enhances endothelial nitric oxide synthase activity by increasing intracellular tetrahydrobiopterin. The Journal of Biological Chemistry. 2000;**275**:17399-17406

[18] www.plantstress.com/articles/oxidative%20stresshtm

[19] Gershoff SN. Vitamin C (ascorbic acid): New roles, new requirements? Nutrition Reviews. 1993;**51**:313-326

[20] Locato V, Cimini S, De Gara L. Strategies to increase vitamin C in plants: From plant defense perspective to food biofortification. Frontiers in Plant Science. 2013;**4**:152. DOI: 10.3389/fpls.2013.00152.

[21] Rose RC, Bode AM. Biology of free-radical scavengers-an evaluation of ascorbate. FASEB Journal. 1993;7:1135-1142

[22] Sato K, Niki E. Shimasaki H. Free radical-mediated chain oxidation of low density lipoprotein and its synergistic inhibition by vitamin E and vitamin C. Archives of Biochemistry and Biophysics. 1990;**279**:402-405

[23] Halliwell B, Gutteridge JM. Free Radicals in Biology and Medicine. New York: Oxford University Press; 1999

[24] https://openi.nlm.nih.gov/detailedresult.php?img=PMC4085347_gr1&req=4

[25] Buettner G R. The pecking order of free radicals and antioxidants: Lipid peroxidation, α-tocopherol, and ascorbate. Archives of Biochemistry and Biophysics. 1993;**300**:535-543

[26] Carr A, Frei B. Does vitamin C act as a pro-oxidant under physiological conditions? FASEB Journal. 1999;**13**(9):1007-1024

[27] Herbert V. Antioxidants, pro-oxidants, and their effects. JAMA. 1994;272:1659-1660

[28] Griffiths HR, Lunec J. Ascorbic acid in the 21st century—More than a simple antioxidant. Environmental Toxicology and Pharmacology. 2001;**10**:173-182

[29] Steinmetz KA Potter JD. Vegetables, fruit and cancer prevention: A review. Journal of the American Dietetic Association. 1996;**96**:1027-1039

[30] Rumsey SC, Levine M. Absorption, transport, and disposition of ascorbic acid in humans. Nutritional Biochemistry. 1998;**9**:116-130

[31] http://www.academia.edu/32451883/AN_OVERVIEW_OF_ASCORBIC_ACID_BIOCHEMISTRY_ASKORB%C4%90K_AS%C4%90T_B%C4%90YOK%C4%90MYASINA_B%C4%90R_BAKI%C5%9E

[32] https://searchenginereports.net/en.wikipedia.org/wiki/Pro-oxidant

[33] https://en.wikipedia.org/wiki/Ascorbic_acid_(molecular_aspects)

[34] https://searchenginereports.net/www.netsci-journal.com/97v4/97014/vitc7.html

Chapter 3

Radioprotective Effect of Vitamin C as an Antioxidant

Tetsuo Yamamoto and Manabu Kinoshita

Additional information is available at the end of the chapter

http://dx.doi.org/10.5772/intechopen.68988

Abstract

Vitamin C is known as a potent antioxidant. We studied vitamin C as a radioprotective agent, focusing on its antioxidative effect. When the body is exposed to radiation, free radicals and reactive oxygen species (ROS) are produced and oxidize cell components, resulting in cell damage. Vitamin C has the potential to scavenge these radical products, thereby protecting against radiation-induced cell damage. We investigated the effects of vitamin C on radiation-induced gastrointestinal (GI) syndrome in mice. The mice received whole-body irradiation followed by bone marrow transplantation 24 h after exposure. Despite avoiding bone marrow failure, the mice eventually died of GI syndrome. Pretreatment with *per os* administration of high-dose vitamin C effectively mitigated radiation-induced GI syndrome and improved mouse survivals, while per os post-treatment with vitamin C was ineffective, presumably due to impaired absorption from the radiation-damaged intestine. We also investigated the effect of post-exposure treatment with intraperitoneal administration of vitamin C on radiation-induced bone marrow dysfunction in mice. Intraperitoneal administration with high-dose vitamin C, even at 24 h after whole-body irradiation, was still effective in avoiding bone marrow dysfunction, thereby increasing mouse survival after radiation. In conclusion, administration of high-dose vitamin C effectively reduced the radiation lethality in mice.

Keywords: antioxidant, radioprotectant, acute radiation syndrome, gastrointestinal syndrome, hematopoietic syndrome

1. Introduction

Since the discovery of radiation at the end of nineteenth century, radiation has been used in various fields, such as medicine, industrials, and agriculture among others. While such efforts have dramatically improved our level of living, radiation has also unfortunately been used to develop nuclear weapons. In addition, accidents have occurred at nuclear power plants that have resulted in catastrophic disasters [1–5]. Because of the accident at Fukushima Dai-ichi Nuclear Power Plant in Japan, many residents near the nuclear plant had to be evacuated and still cannot return to their homes [6].

Ionizing radiation affects the human body in two ways. It affects the body directly by cutting DNA chains, resulting in cell injuries. To reduce the direct radiation damage, lead is used to protect from gamma rays, and water or boron is also used to protect from neutrons. Physical protection is essential to prevent the direct effects of radiation. It also affects the body indirectly, as when radiation hits water molecules in host cells, it produces large amounts of free radicals and reactive oxygen species (ROS), which then oxidize the cell components, resulting in cell injuries [7–10] (**Figure 1**).

To reduce/mitigate the indirect effects of radiation, it is essential to scavenge the generated free radicals or ROS [11, 12]. As host cells have abundant antioxidative enzymes by nature, free radicals and ROS generated under normal conditions are easily scavenged. However, if a body is exposed to radiation and exaggerated amounts of free radicals and ROS are produced, they cannot be entirely scavenged by intrinsic antioxidative enzymes. In such a situation, the administration of an antioxidative product can facilitate scavenging and prevent cell damage due to radiation. Many antioxidative products are reported to have radioprotective effects [13–17]. In this section, we describe the radioprotective effects of vitamin C, which is one of the strongest antioxidative agents [18–20] (**Figure 1**).

Vitamin C was originally discovered as a medical treatment for scurvy. It has several pharmacological actions, including an antioxidative effect. Nowadays, vitamin C is widely used as an additive in many foods. Although most animals can synthetize vitamin C by themselves, some primates, including human have lost the ability to produce vitamin C and so need to consume it in some form [21, 22].

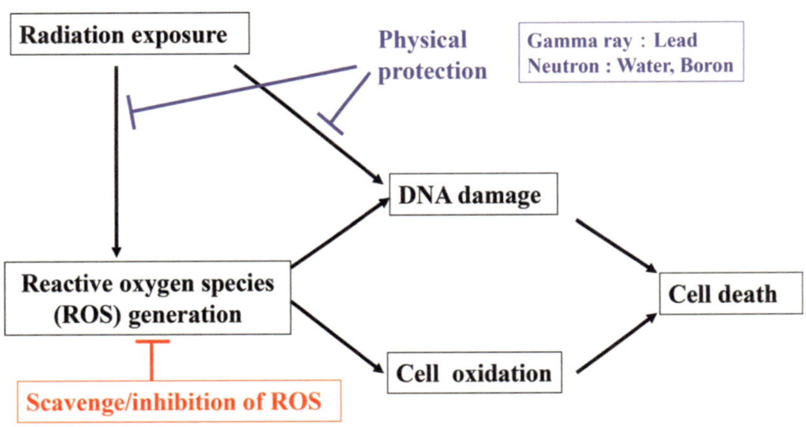

Figure 1. Our strategy for the protection against radiation injury.

Acute radiation syndrome occurs when a host is exposed to high-dose radiation in a short period of time. This syndrome can develop even in animals (e.g., mice) that can synthetize vitamin C by themselves, as deficiency of antioxidants necessarily occurs after massive radiation exposure. Hematological syndrome, gastrointestinal (GI) syndrome, and demagogical syndrome are highly frequent symptoms. The pathophysiology of acute radiation syndrome is basically caused by a stem cell disorder, which has a short cycle of cell turn over [23, 24].

Our group investigated the antioxidative effect of vitamin C on irradiated hosts using a mouse model.

2. Effects of pretreatment with *per os* (p.o.) administration of vitamin C on GI syndrome in whole-body-irradiated mice

We investigated the effect of vitamin C on radiation-induced GI syndrome in mice receiving whole-body irradiation (WBI). We found that pretreatment with vitamin C significantly improved the GI syndrome, thereby rescuing mice that had undergone bone marrow transplantation from lethal radiation exposure [18].

2.1. Bone marrow transplantation alone cannot rescue mice from WBI at 14 Grays

When the mice were exposed to WBI at 6–14 Gray (Gy), 80% of mice survived after WBI at 6 Gy, but no mice survived at doses of ≥8 Grays (Gy) (**Figure 2A**). However, when the mice received bone marrow transplantation (BMT) at 24 h after WBI, their mortality were drastically reduced and most mice survived (8 and 10 Gy, 100% survival; 12 Gy, 75% survival). After exposure of WBI at 8–12 Gy, the mice presumably died of bone marrow dysfunction, namely hematopoietic syndrome, and thereby BMT was effective in these mice. Nevertheless, BMT was not effective in the mice after WBI at 14 Gy (**Figure 2B**).

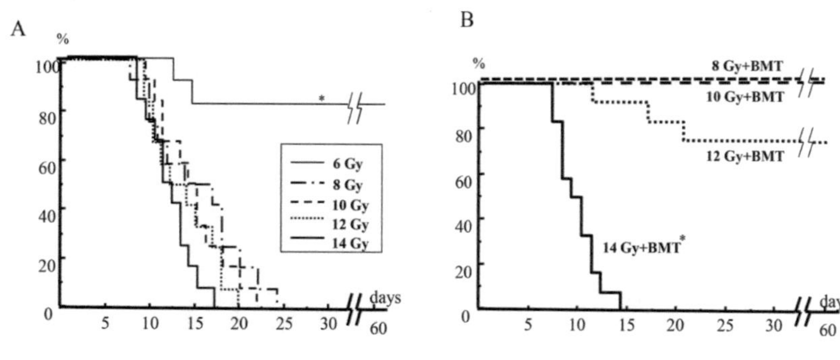

Figure 2. The mouse survival after WBI (A) and the effect of BMT on the survival after WBI (B) [18].

2.2. Whole-body-irradiated mice at 14 Gy showed marked denudation of the intestinal mucosa and bone marrow aplasia

By the pathological examination, severe bone marrow aplasia was observed in mice at 7 days after WBI at 8 Gy compared to unirradiated control mice (0 Gy) (**Figures 3A-a, B-a**), although their intestinal mucosae were still intact even after 8 Gy radiation exposure (**Figures 3A-b, c, B-b, c**). However, when the mice were expose to WBI at 14 Gy, marked denudation of the intestinal mucosae were observed in them at 7 days after irradiation (**Figures 3C-b, c**, arrowheads) accompanied with severe bone marrow aplasia (**Figure 3C-a**). These mice suffered from the radiation-induced GI damage and presumably died of this GI damage after WBI at 14 Gy, despite rescuing bone marrow failure by BMT.

2.3. Pretreatment but not posttreatment with vitamin C improved the mouse survival after 14-Gy WBI, in combination with BMT

To examine the dose response effects of vitamin C on the irradiated mice, we p.o. administered 1.5, 15, 150 and 1500 mg/kg/day of vitamin C for 3 days before 14-Gy WBI to mice, followed by BMT at 24 h after radiation. Pretreatment with 150 mg/kg/day of vitamin C was able to rescue some subjected mice (42% survival), while other doses of pretreatment with vitamin C

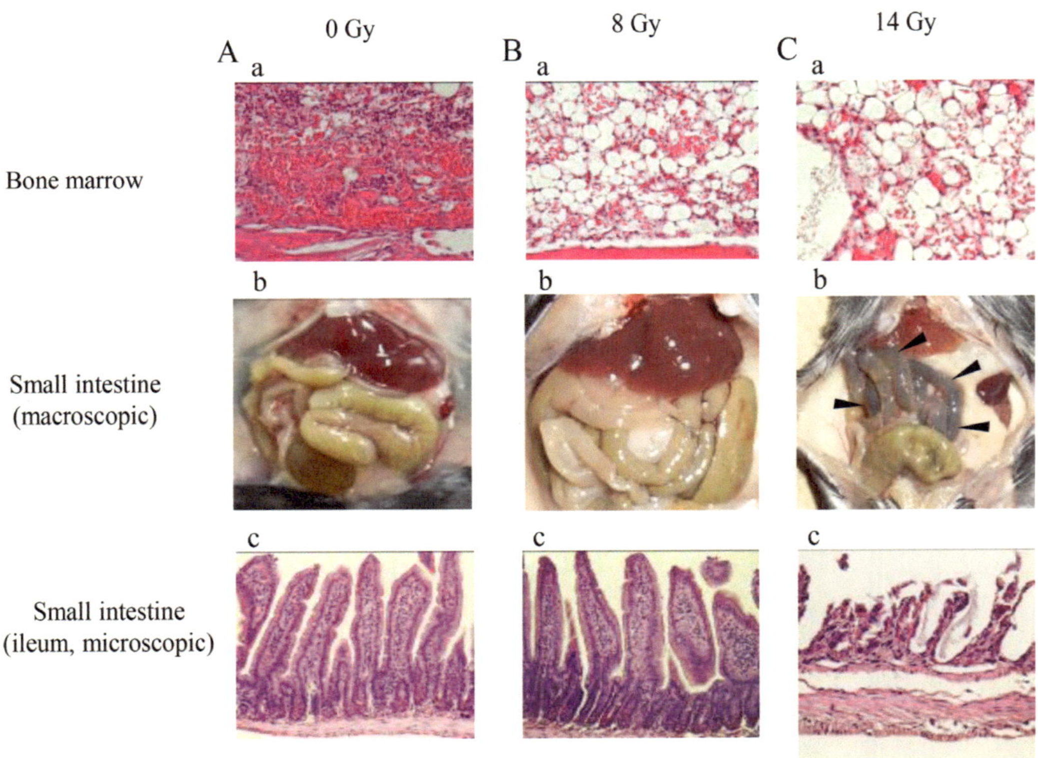

Figure 3. Radiation-induced damage to the bone marrow and small intestine in mice 7 days after radiation [18].

rescued no mice. Pretreatment with 1500 mg/kg/day of vitamin C also did not improve the mouse survival after WBI at 14 Gy. Such a massive administration of vitamin C may be conversely harmful for the host. We next studied the effect of posttreatment with vitamin C on the irradiated mice. Mice were p.o. administered with 150 mg/kg/day of vitamin C for 3 days after 14-Gy WBI and received BMT. Posttreatment with vitamin C did not affect the survival (0% survival) (**Figure 4**), although the pretreatment with the same doses of vitamin C was effective. When the mice were only pretreated with vitamin C and did not receive BMT, no mice survived after 14-Gy WBI (**Figure 4**). Of note: these mice died of bone marrow aplasia after radiation, not GI damage.

2.4. Pretreatment with vitamin C markedly improved the radiation-induced intestinal damage, thereby rescuing mice from lethal WBI at 14 Gy, in combination with BMT

BMT following 14-Gy WBI remarkably improved bone marrow aplasia in the mice 7 days after radiation. However, these mice showed severe degenerative changes in the intestinal mucosa (**Figures 5A-a, b**). Pretreated with vitamin C for 3 days before 14-Gy WBI without BMT markedly improved the mucosal degeneration in the intestine but did not improve the bone marrow aplasia in mice (**Figures 5B-a, b**). Notably, when mice were pretreated with vitamin C and received BMT following 14-Gy WBI, marked improvements in both bone marrow aplasia and intestinal mucosal degeneration were noted (**Figure 5C-b**).

2.5. Pretreatment with vitamin C prevented the intestinal tissue damage of the mice receiving 14-Gy WBI followed by BMT

The mice showed significantly lower villus heights and crypt counts per circumference in the intestine after 14-Gy WBI (but not those receiving 8-Gy WBI) (**Figure 6A, B**). Although

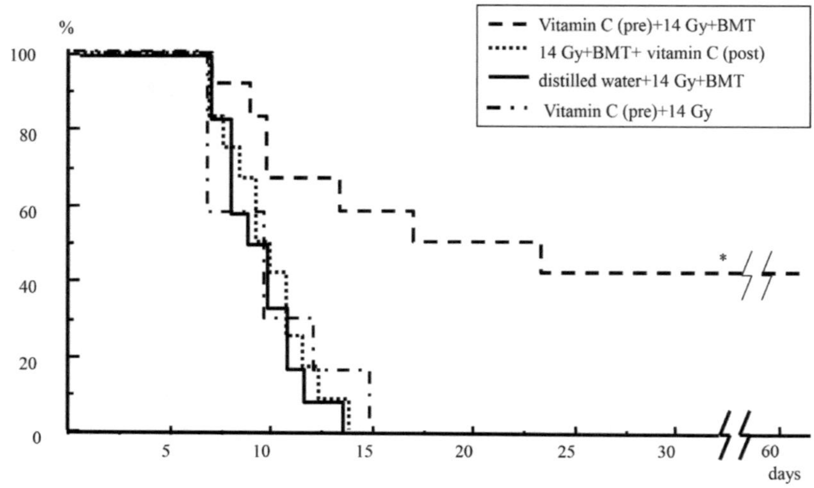

Figure 4. The effect of pre and posttreatment with vitamin C on the mouse survival after WBI [18].

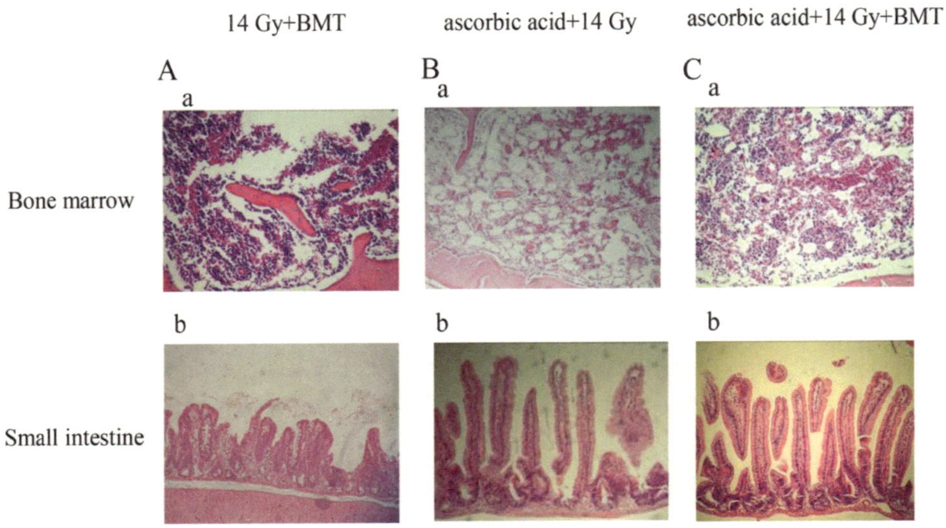

Figure 5. The effect of pretreatment with vitamin C or BMT following WBI on the bone marrow or small intestine in irradiated mice [18].

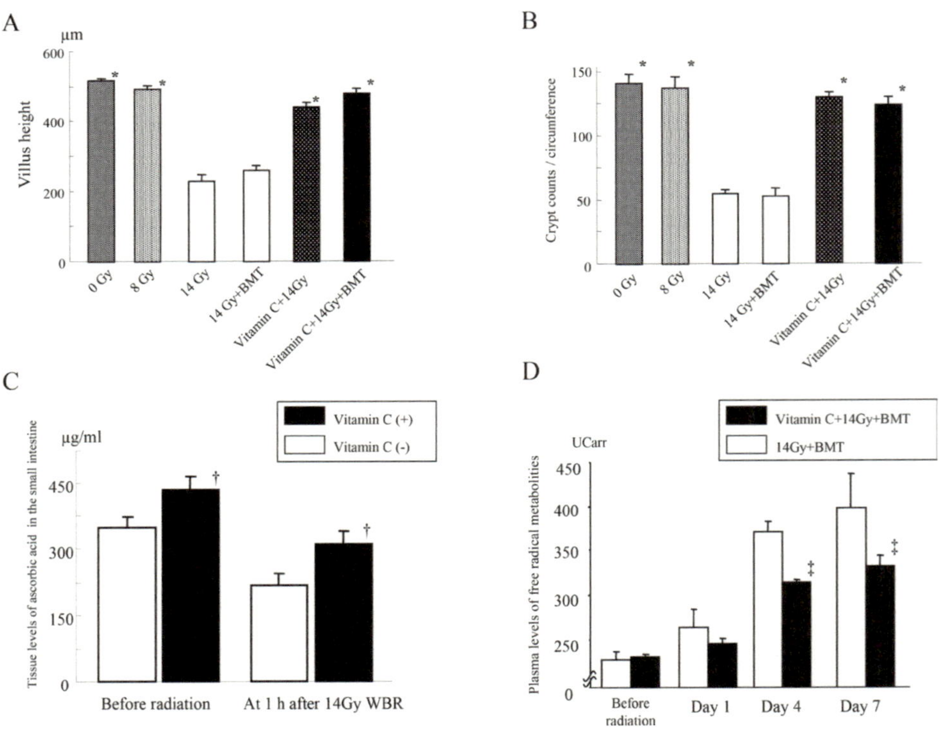

Figure 6. Villus height (A), and crypt counts (B) of the small intestine in mice. The tissue vitamin C levels in the small intestine (C), and the production of free radical metabolites in mice irradiated at 14 Gy (D) [18].

treatment with BMT alone following 14-Gy WBI did not affect these degenerative changes, pretreatment with vitamin C markedly improved them (**Figure 6A, B**). Both the mice that were pretreated with vitamin C and received BMT following 14-Gy WBI also showed significant improvements (**Figure 6A, B**).

2.6. Oral administration of vitamin C increased the tissue concentration of vitamin C in the small intestine

Per os administration of vitamin C for 3 days significantly increased the plasma concentrations of vitamin C in the treated mice than that of untreated controls (57 ± 11 versus 30 ± 8 μ/mL, $P < 0.05$). Next, the vitamin C levels in tissues of small intestine were examined. To measure vitamin C levels in tissue of small intestine, a sample of small intestine (0.7 g) was removed from each mouse immediately after sacrifice and homogenized in 5.4% metaphosphoric acid (9.8 g). Vitamin C levels of the homogenate supernatant were measured in the SRL laboratory (Tokyo, Japan) using high performance liquid chromatography (HPLC). Pretreatment with vitamin C significantly increased the tissue concentrations of vitamin C in the small intestine just before radiation (**Figure 6C**). Interestingly, the tissue vitamin C levels were decreased at 1 h after radiation in not only vitamin C-pretreated mice but also in untreated control mice (**Figure 6C**), indicating the critical consumption of tissue vitamin C by irradiation. Nevertheless, the mice pretreated with vitamin C still showed significantly higher tissue vitamin C levels than the untreated control mice (**Figure 6C**).

2.7. Pretreatment with vitamin C suppressed a radiation-induced increase in the free radical metabolites in mouse plasma

To measure free radical metabolites in the plasma, we used the d-ROMs test (Diacron, Grosseto, Italy). It is a spectrophotometric method that assesses overall oxidative stress by measuring total hydroperoxide levels, given that hydroperoxides are intermediate oxidative products of lipids, peptides, and amino acids. We diluted 0.02 mL plasma in 1 mL acetate-buffered solution. Hydroperoxide groups react with the transition metal ions liberated from the proteins in the acidic medium, and are converted to alkoxyl and peroxyl radicals according to the Fenton reaction. These newly formed radicals, the quantities of which are directly proportional to those of the peroxides, were trapped chemically with 0.02 mL chromogen (N,N-diethyl-para-phenylenediamine), leading to the formation of a radical cation of this chromogen. The purple color resulting from this reaction over time was monitored in a spectrophotometer (Wismarll FRAS4, Tokyo, Japan) at 505 nm. The results of this method were expressed in conventional units (Carratelli units [UCarr]). Although free radical metabolites gradually increased in the plasma of untreated mice after radiation, pretreatment with vitamin C significantly suppressed the increase in free radical metabolites at 4 and 7 days after radiation (**Figure 6D**). Posttreatment with vitamin C did not suppress the increase in free radical metabolites in mice.

2.8. Pretreatment with vitamin C suppressed the radiation-induced DNA damage in the crypt epithelial cells of the mouse's small intestine

To examine the effect of pretreatment with vitamin C on the DNA damage in the mouse's small intestine after WBI, antisingle stranded (ss) DNA was stained in the samples of small intestine, using polyclonal rabbit anti-ssDNA (A4506, DAKO, Glostrup, Denmark). In the mice without vitamin C pretreatment, the number of positive-stained cells for ssDNA increased in the epithelial crypts of the small intestine at 6 h after radiation and further increased at 24 h (**Figure 7**). In contrast, pretreatment with vitamin C significantly suppressed the increase in ssDNA positive-stained cells (**Figure 7**). In these mice, the epithelial cells in the crypts of small intestines only showed slightly positive staining for ssDNA at 24 h after radiation (**Figure 7**). Radiation-induced DNA damage of the mouse intestinal crypt cells may be effectively inhibited by pretreatment with vitamin C.

3. The drastic effect of combination therapy with p.o. administration of vitamin C on the GI syndrome in mice receiving abdominal radiation

Although the survival rates of the mice receiving WBI at 14 Gy followed by BMT were increased by pretreatment with vitamin C, more than half of the mice still died of radiation-induced GI damage. We therefore modified the administration manner of vitamin C in order to augment its radioprotective potential.

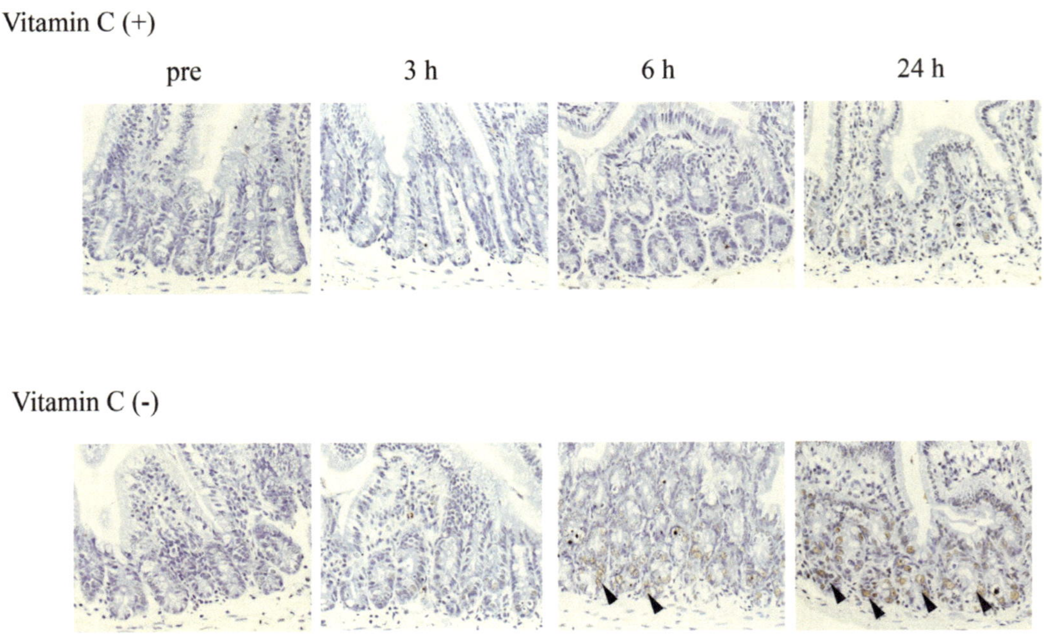

Figure 7. Radiation-induced DNA damage in the small intestines of mice irradiated with 14 Gy [18].

We used an abdominal radiation model that was not complicated with bone marrow damage. Therefore, this model does not require murine sacrifice as a donor of bone marrow cells and is able to simplify the effects of the injury and treatment. In addition, abdominal radiation is frequently performed on patients suffering from abdominal malignancy. It is therefore rational to research and develop an effective therapy for radiation-induced GI damage using a model of abdominal radiation. Although posttreatment with vitamin C alone was ineffective in our previous study, combined therapy of pre and posttreatment with vitamin C improved the survival rates slightly after abdominal radiation. We also added one-shot engulfment (boosting) of vitamin C 8 h before radiation in order to effectively increase the tissue vitamin C levels at the time of radiation exposure. As a result, we were able to elicit a remarkable radioprotective effect against radiation-induced lethal GI damage (100% survival) by combination therapy.

3.1. Abdominal radiation at 10–12 Gy was not lethal for mice, while the same radiation doses of WBI were lethal

As we described above, no mice survived after WBI at doses of ≥8 Gy (**Figure 2A**). In contrast, all mice survived after abdominal radiation even at 10 Gy, but no mice survived after abdominal radiation at ≥13 Gy (**Figure 8**).

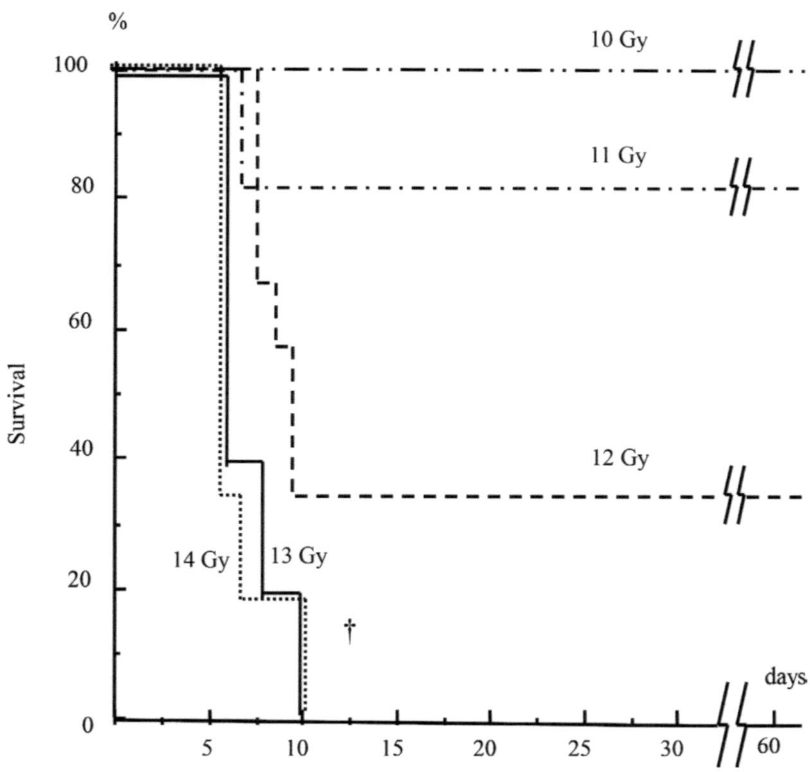

Figure 8. The survival of mice after abdominal radiation [19].

3.2. Abdominal radiation at 13 Gy caused lethal GI damage without inducing bone marrow aplasia

Severe bone marrow aplasia was observed in the lumbar vertebrae, the sternum, and the femur in mice at 7 days after WBI at 8 Gy (**Figure 9**). In contrast, the mice after abdominal radiation at 13 Gy retained a substantial number of bone marrow cells in the sternum and femur but not in the lumbar vertebrae that was directly exposed to a substantial dose during abdominal radiation (**Figure 9**). Bone marrow function may persist even after lethal abdominal radiation (13 Gy). However, marked denudation of the gastrointestinal mucosa, especially the ileac mucosa, was observed in the mice at 7 days after abdominal radiation at 13 Gy (**Figure 9**). In contrast, the mice receiving lethal WBI at 8 Gy did not show such severe intestinal damage (**Figure 9**). Abdominal radiation thus induced severe GI damage but not extensive bone marrow damage in mice.

3.3. Abdominal radiation at 13 Gy restored the white blood cell (WBC) counts but not the plasma citrulline levels in mice

WBC counts transiently decreased at 1–3 days after abdominal radiation at 11–13 Gy but increased in mice beyond 5 days after abdominal radiation (**Figure 10A**). The mice receiving

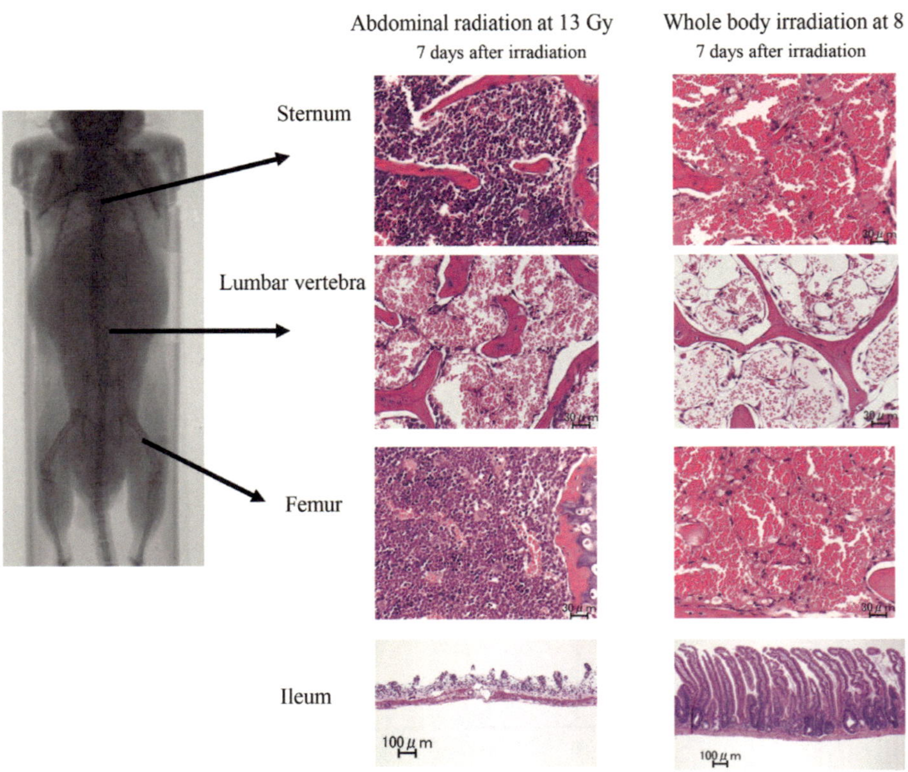

Figure 9. The pathological findings of the bone marrow and ileum in mice after abdominal radiation at 13 Gy or WBI at 8 Gy [19].

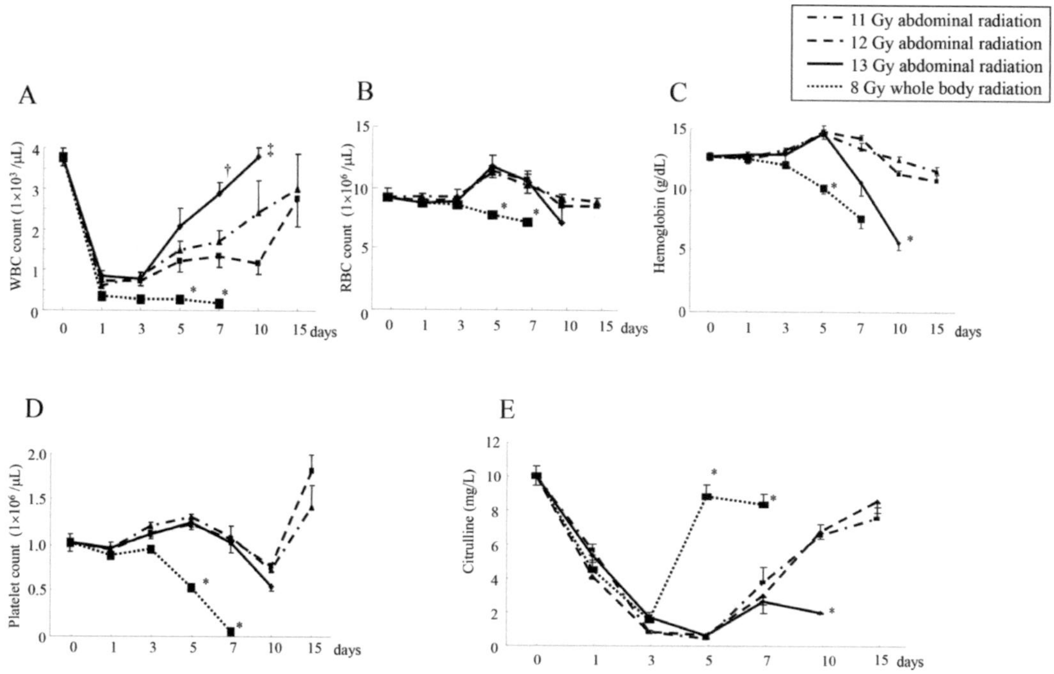

Figure 10. The changes in the hematological parameters and plasma citrulline levels after abdominal radiation [19].

WBI at 8 Gy did not show such a restoration of the WBC counts and eventually died of irreversible lethal bone marrow damage (**Figure 10A**). The red blood cell (RBC) counts as well as the hemoglobin (Hb) levels were also increased around 5–7 days after abdominal radiation at 11 and 12 Gy (**Figure 10B, C**), whereas such increases in the RBC counts or Hb levels were not observed in the mice after 8-Gy WBI (**Figure 10B, C**). Interestingly, the mice showed severe anemia, as assessed by the RBC counts and Hb levels, at 10 days after abdominal radiation at 13 Gy due to gastrointestinal bleeding and subsequently died (**Figure 10B, C**). The platelet counts did not obviously decrease after abdominal radiation at 11–13 Gy, while in contrast, those counts markedly decrease after WBI at 8 Gy (**Figure 10D**). Abdominal radiation, even at 13 Gy (which is lethal), may not cause severe suppression of the bone marrow cells in mice. Plasma citrulline level was measured using a fully automated amino acid analyzer (JLC-500/V2, Nihon Denshi, Tokyo, Japan). It can be reportedly utilized as a biomarker of intestinal failure after massive resection of the small intestine. Therefore, the changes in the plasma levels may be an effective biomarker of radiation-induced GI syndrome [19]. Although the plasma citrulline levels were decreased in mice around 3–7 days after abdominal radiation at 11 and 12 Gy, they recovered 10 days after radiation exposure (**Figure 10E**). However, the mice receiving abdominal radiation at 13 Gy did not show restoration of the citrulline levels at 10 days (**Figure 10E**). Since their intestinal mucosae were also severely impaired at that point (**Figure 9**), the plasma citrulline levels after abdominal radiation may reflect the change in the intestinal degradation. In line with this finding, the mice receiving WBI at 8 Gy showed significant restoration of the plasma citrulline levels at 5–7 days (**Figure 10E**) without any intestinal damage (**Figure 9**).

3.4. Combination therapy with vitamin C drastically improved the mouse survival after abdominal radiation

Per os administration of vitamin C for 3 days before irradiation (Plan I, **Figure 11**) rescued only 20% of mice from lethal abdominal irradiation at 13 Gy (**Figure 12**). Engulfment of vitamin C at 8 h before radiation (Plan II, **Figure 11**) also rescued 20% of mice from abdominal radiation at 13 Gy (**Figure 12**). We tried the engulfment of vitamin C at 2 h before radiation in the mice, but their survival was <10%, suggesting that an interval of 2 h was too short to sufficiently increase the vitamin C levels in the mice. We next examined the p.o. administration of vitamin C for 3 days and engulfment 8 h before radiation in mice (Plan III, **Figure 11**). Nevertheless, the survivals of these mice were still 20% (**Figure 12**). As expectedly, posttreatment with p.o. administration of vitamin C was ineffective (0% survival, **Figure 12**). However interestingly, when we tried combination treatment before and after radiation in mice (Plan V, **Figure 11**), their survival rates increased to 40% (**Figure 12**). Notably, when the one-shot engulfment (8 h before radiation) of vitamin C was added to the oral administration for 10 days before/after radiation (Plan VI, **Figure 11**), the mouse survival was drastically increased to 100% survival after abdominal radiation (**Figure 12**).

3.5. Combination therapy with vitamin C effectively increased the tissue vitamin C levels in the small intestine of mice

Pretreatment with vitamin C for 3 days before radiation significantly increased the vitamin C levels in the intestinal tissue just before radiation, and also the administration of engulfing

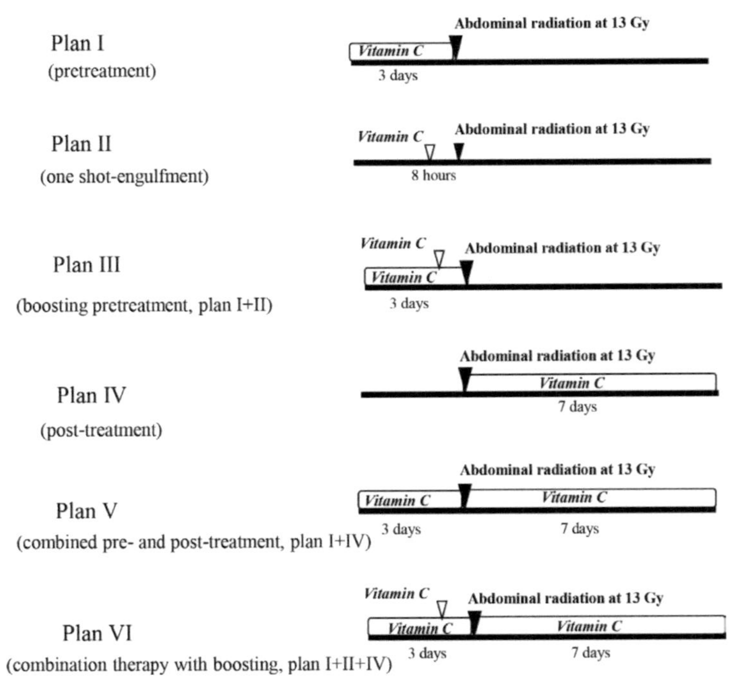

Figure 11. The experimental design for abdominal radiation and the treatments with vitamin C [19].

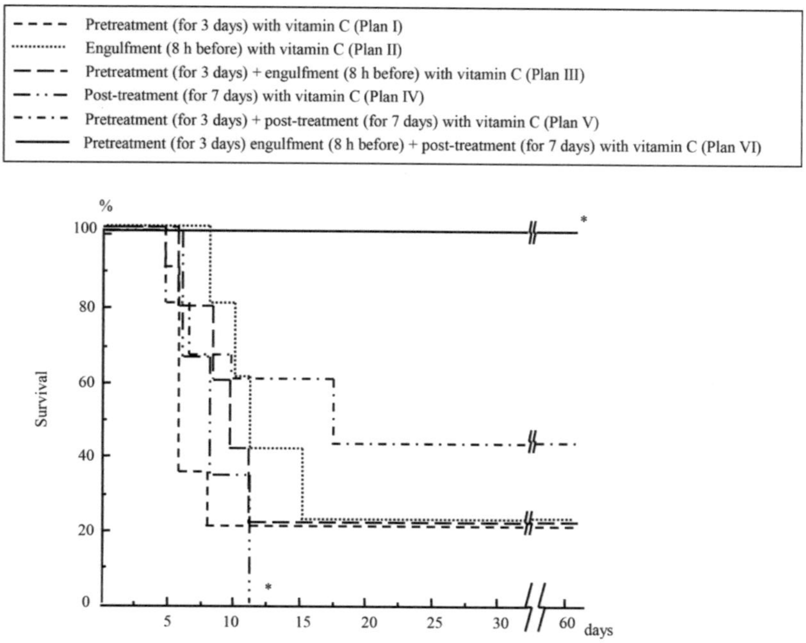

Figure 12. The effects of treatment with vitamin C on the survival of mice receiving abdominal radiation [19].

vitamin C at 8 h before radiation tended to increase the tissue vitamin C levels, but not significantly (**Table 1**). Interestingly, boosted pretreatment with vitamin C (oral intake for 3 days and engulfment at 8 h before radiation) further increased the tissue vitamin C levels in the intestine just before radiation (**Table 1**). However, there were no significant differences in the plasma vitamin C levels among these mouse groups (**Table 1**). Boosted p.o. pretreatment with vitamin C effectively increased the tissue vitamin C levels in the small intestine just before radiation.

3.6. Combination therapy with vitamin C restored the intestinal damage while reducing the elevation of free radical metabolite levels after abdominal radiation in mice

Combination therapy with p.o. administration of vitamin C for 10 days before/after radiation and one-shot engulfment at 8 h before radiation (Plan VI, **Figure 11**) significantly restored the intestinal damage in mice after abdominal radiation at 13 Gy (**Figure 13A**). This combination therapy with vitamin C also suppressed the positive TUNEL (terminal deoxynucleotidyltransferase-mediated dUTP nick end labeling) staining in the ileac mucosa in mice at 12 h after abdominal radiation at 13 Gy (**Figure 13B**). TUNEL staining was performed using an *in-situ* apoptosis detection kit (MK500, Takara, Tokyo Japan). Vitamin C may suppress the radiation-induced apoptosis in the intestinal mucosa. Consistently, combination therapy with vitamin C also restored the villus height of the ileac mucosa beyond 7 days after abdominal radiation (**Figure 14A**) and restored the plasma citrulline levels (**Figure 14B**). Thus, a certain relationship was suggested between the plasma citrulline levels and intestinal damage. This regiment of vitamin C treatment drastically improved the radiation-induced intestinal damage, resulting in an improvement in mouse survival after lethal abdominal radiation

	Vitamin C levels	
Groups	**Tissue level of the small intestine (mg/L)**	**Plasma (mg/L)**
No treatment with vitamin C	22.8 ± 2.0	5.2 ± 0.5
Engulfment (8 h before) with vitamin C	27.1 ± 1.4	5.2 ± 0.9
Pretreatment (for 3 days) with vitamin C	30.5 ± 1.6 †	6.2 ± 0.7
Pretreatment (for 3days) + engulfment (8 h before) with vitamin C	36.1 ± 1.7 *	6.3 ± 0.6

Mice received the indicated treatments with vitamin C. The small intestine and plasma were obtained from the mice just before radiation. Data are shown as means ± SE (standard error) from n=5 in each group. *$p<0.01$ vs. no treatment and engulfment (8h) and $p<0.05$ vs. pretreatment (3 days), and †$p<0.01$ vs. no treatment [19].

Table 1. Vitamin C levels in the small intestine and plasma after the treatments with vitamin C.

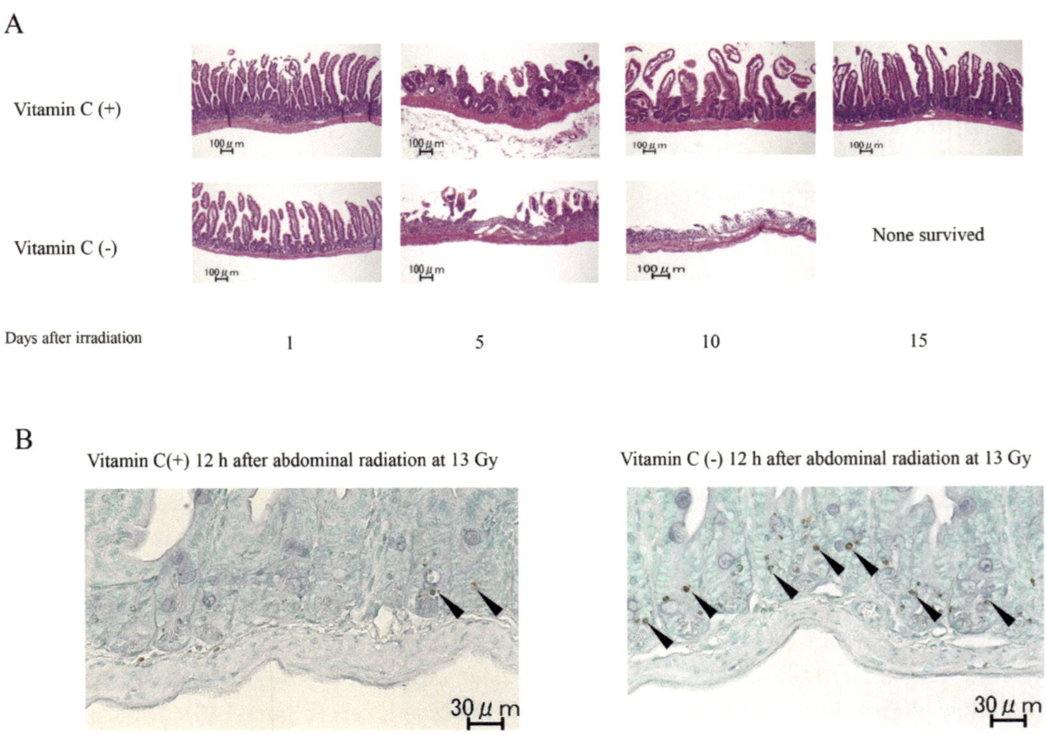

Figure 13. The changes in the intestinal mucosa of the irradiated mice with or without vitamin C treatment [19].

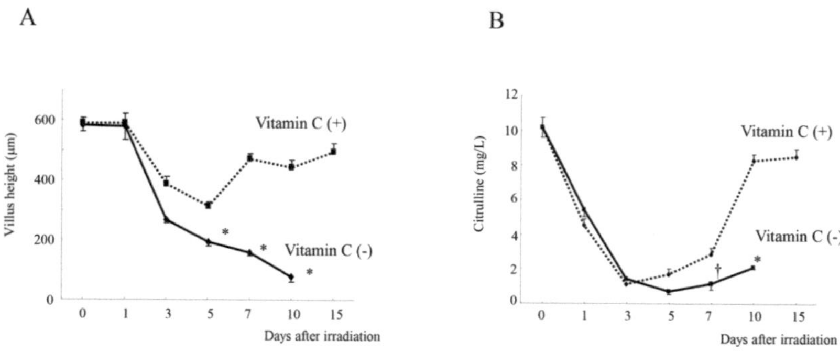

Figure 14. The changes in the villus height and plasma citrulline levels in the irradiated **Figure 14**. mice with and without vitamin C treatment [19].

4. The effect of postexposure treatment with intraperitoneal administration of vitamin C on the bone marrow dysfunction in mice after WBI

Considering the practical applications of vitamin C, such as in treating those affected by nuclear accidents, postexposure treatment is crucial for patients suffering from radiation exposure. However, it can be difficult for the patients to take high-dose vitamin C orally after radiation exposure, because the GI tract may already be damaged by radiation. Intravenous (or intraperitoneal; in case of mice) administration is preferred.

We therefore investigated the effects of postexposure treatment with intraperitoneal (i.p.) vitamin C on mice after irradiation. We found that the postexposure treatment with vitamin C reduced radiation-induced apoptosis in bone marrow cells and restored the hematopoietic function, thereby reducing the mortality in irradiated mice. Interestingly, postexposure treatment with vitamin C was effective even 24 h after radiation exposure. Large amounts of vitamin C (3 g/kg) were needed to improve the radiation-induced mortality in mice. We next examined the effects of divided i.p. administration of high-dose vitamin C. Divided i.p. administration with vitamin C (1.5 g/kg × 2, immediately after and 24 h after radiation) was effective in treating irradiated mice. The administration of high-dose vitamin C may be useful for mitigation therapy even after exposure [20].

4.1. Postexposure treatment with i.p. administration of vitamin C improved the survival of whole-body-irradiated mice

When the mice were exposed to 7 Gy-WBI, their survival rate was 67%. However, postexposure treatment with 3 g/kg of vitamin C immediately after WBI at 7 Gy rescued all of the subject mice (100% survival) (**Figure 15A**). Nevertheless, either 1 or 2 g/kg of vitamin C was ineffective (**Figure 15A**). A substantial dose of vitamin C (3 g/kg) was needed to induce a mitigating effect against radiation-induced lethality. This postexposure treatment with 3 g/kg of vitamin C

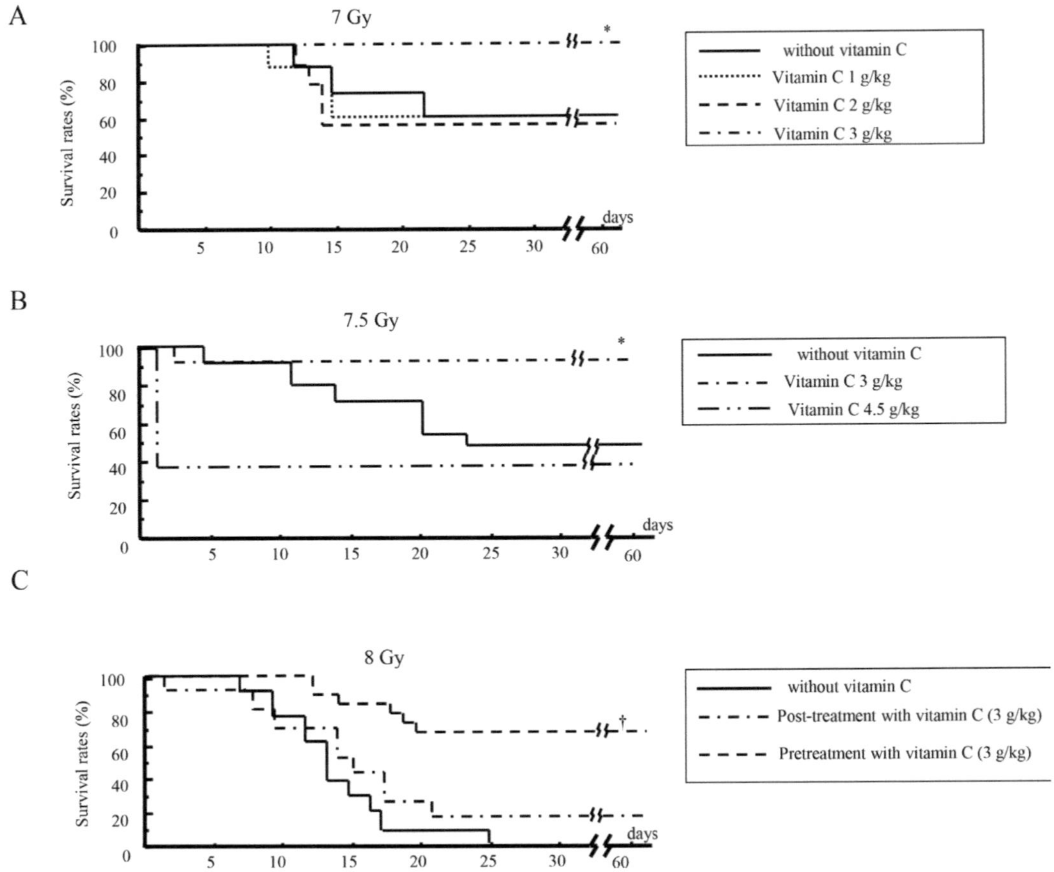

Figure 15. The survival of mice that received postexposure treatment with vitamin C after WBI at 7 (A), 7.5 (B), and 8 Gy (C) [20].

also significantly improved the mouse survival after 7.5-Gy WBI (**Figure 15B**). However, treatment with 4.5 g/kg of vitamin C conversely reduced the mouse survival after 7.5-Gy WBI (**Figure 15B**). With this treatment, 60% of mice died within 1 day after vitamin C treatment following radiation. The massive administration of vitamin C may be harmful for these irradiated mice, as more than half of the mice died within 1 day of the administration of 4.5 g/kg vitamin C, even in the absence of irradiation (**Figure 15B**). As expectedly, pretreatment with 3 g/kg of vitamin C rescued 65% of mice after lethal 8-Gy WBI, although the postexposure treatment with vitamin C (3 g/kg) rescued 20% of mice (**Figure 15C**), confirming that pretreatment with vitamin C had a potent radioprotective effect. Postexposure treatment with 4 g/kg of vitamin C as well as 4.5 g/kg markedly reduced the mouse survival at 1 day after 8-Gy WBI (13 and 0%, respectively), suggesting a harmful effect due to the extremely high-dose of vitamin C.

4.2. Postexposure treatment with vitamin C restored the bone marrow functions in mice after WBI

Postexposure treatment with 3 g/kg of vitamin C did not restore the WBC counts, RBC counts, Hb levels, or platelet counts in mice until 14 days after WBI at 7.5 or 8 Gy, but markedly restored these hematological parameters at 3 weeks after WBI (**Figure 16**). Lethal 8-Gy WBI had severely damaged bone marrow cells at 14 days after radiation (**Figure 17A, C**). However, postexposure treatment with 3 g/kg vitamin C rescued a portion of the bone marrow cells from the lethal radiation at 14 days after exposure (**Figure 17B, D**), despite still suppression of hematological parameters (**Figure 16**, right column). Immunohistochemical staining of caspase-3 was performed using polyclonal rabbit anticaspase-3 antibody (Asp175, Cell Signaling Technology, Inc. Danvers, MA). Caspase-3-positive cells increased in the bone marrow 6 h after WBI at 8 Gy (**Figure 17E**, indicated by arrows). However, postexposure treatment with 3 g/kg of vitamin C reduced the caspase-3 positive cells (**Figure 17F**). Postexposure treatment with vitamin C may suppress radiation-induced apoptotic cell death in the bone marrow.

4.3. Postexposure treatment of vitamin C increased the plasma vitamin C and biological antioxidant potential (BAP) levels in mice after WBI

Intraperitoneal administration with 3 g/kg vitamin C markedly increased the plasma vitamin C levels in mice with and without 7.5-Gy WBI at 30 min and 1 h after administration (**Table 2**), and the levels were then decreased at 2 h (irradiated mice, 120 ± 38; nonirradiated mice, 188 ± 27 μg/L at 2 h). The vitamin C levels in the plasma were measured by the SRL laboratory (Tokyo, Japan) using HPLC. The ferric-reducing ability in plasma was also measured spectrophotometrically using the BAP test (Diacron, Grosseto, Italy). In brief, a colored solution containing ferric ions is reduced to ferrous ions by reduction of the sample, and the antioxidant activity of the sample is proportional to the measured decrease in absorbance. The BAP assays were performed on the FRAS 4 analyzer (Wismarll FRAS 4, Tokyo, Japan) according to the manufacturer's protocols.

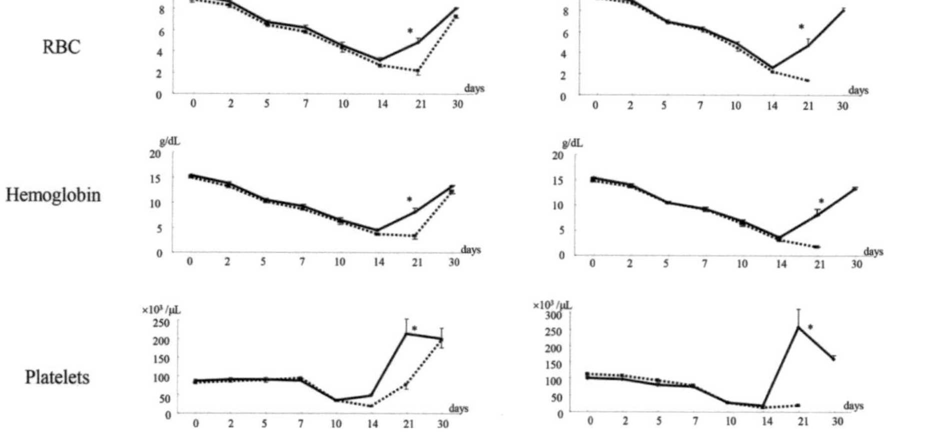

Figure 16. Changes in hematological parameters after irradiation of mice [20].

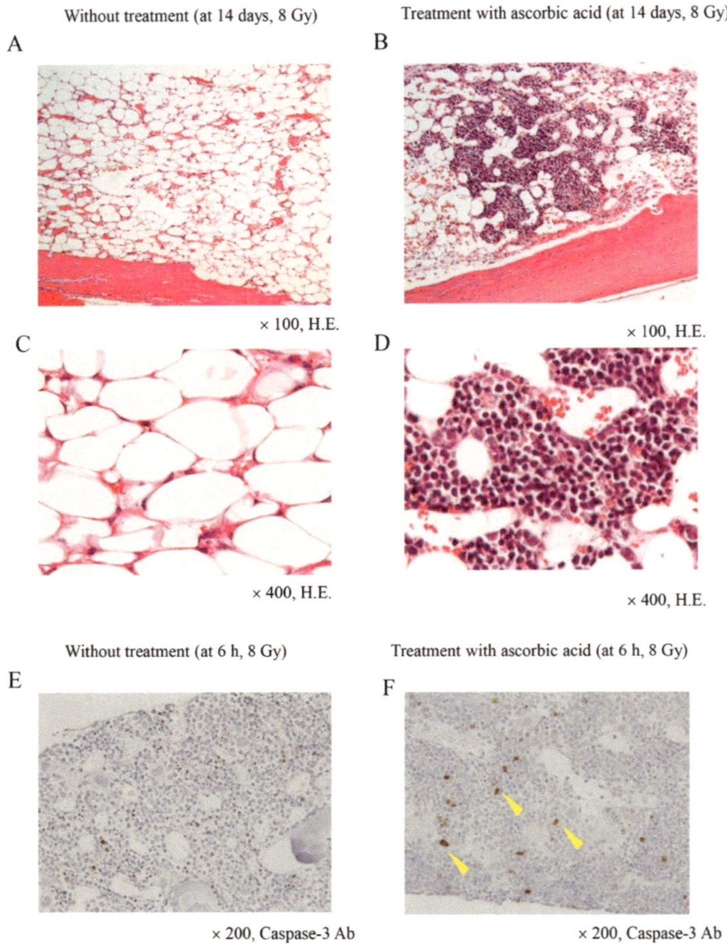

Figure 17. Histological findings in the bone marrow of irradiated mice [20].

Radiation	Treatment	Plasma vitamin C levels (μg/L)		Plasma BAP levels (mMol/L)	
		30 min after administration	1 h after administration	30 min after administration	1 h after administration
7.5 Gy	Vitamin C (+)	3659 ± 382*	1,878 ± 419*	30.1 ± 2.1*	5.7 ± 0.8*
	Vitamin C (−)	2.6 ± 0.7	1.7 ± 0.2	3.1 ± 0.1	3.4 ± 0.1
Nonradiation	Vitamin C (+)	3287 ± 520*	940 ± 128	28.3 ± 2.4*	5.3 ± 0.8
	Vitamin C (−)	1.2 ± 0.2		3.0 ± 0.1	

Mice received WBI at 7.5 Gy or did not. Subsequently, 3 g/kg of vitamin C or saline was administered to the mice. Data shown are mean ± SE from five mice.

* $p < 0.01$ vs. vitamin C (−) [20].

Table 2. Plasma levels of vitamin C after radiation with or without vitamin C treatment.

These plasma BAP levels were markedly increased at 30 min after i.p. administration with vitamin C in both irradiated and nonirradiated mice (**Table 2**). Intraperitoneal administration of vitamin C potently induced antioxidant capability in mice, even after irradiation. Whereas, these increased plasma BAP levels reduced to normal levels in both irradiated and nonirradiated mice at 2 h after i.p. administration of vitamin C (**Table 2**).

4.4. Postexposure treatment of vitamin C at 24 h after radiation was still effective in irradiated mice

To examine how long we could delay postexposure treatment with vitamin C after radiation exposure, vitamin C was administered to mice at 1, 6, 12, 24, 36 or 48 h after 7.5-Gy WBI. As a result, treatment with vitamin C up to 24 h after radiation effectively increased the survival of irradiated mice, although the treatment beyond 36 h postirradiation was ineffective (**Figure 18B**).

4.5. Divided i.p. administration with vitamin C (1.5 g/kg × 2, immediately after and 24 h after radiation) was effective in irradiated mice

Considering a clinical application of vitamin C, a single administration with 3 g/kg of vitamin C appeared to be too high a dose for mice to tolerate, although nonirradiated mice did not die after this dose. The efficacy of two treatments of vitamin C was then examined (1.5 g/kg × 2, immediately after radiation and 24 h after, 3 g/kg in total). This divided i.p. administration with vitamin C markedly increased the survival of mice after WBI at 7.5 Gy (**Figure 19A**),

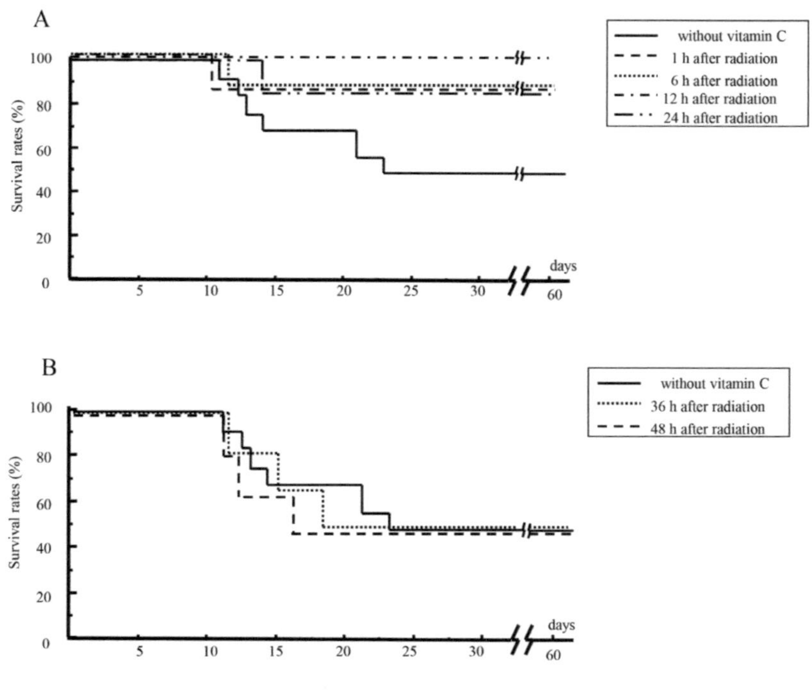

Figure 18. The mouse survival as a function of the time of postexposure treatment with vitamin C [20].

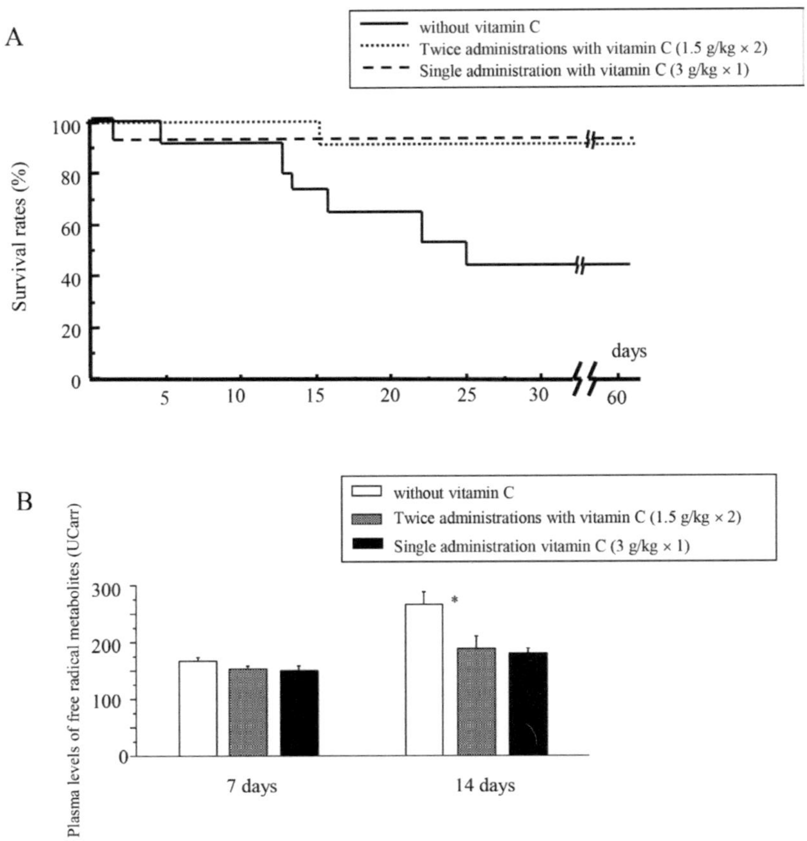

Figure 19. The survival of mice following one or two treatments with vitamin C (3 g/kg) after WBI and, plasma levels of free radical metabolites were also measured 7 and 14 days after WBI [20].

despite either single injection with 1.5 g/kg of vitamin C immediately after or 24 h after radiation was ineffective. In addition, divided i.p. administration with vitamin C (1.5 g/kg × 2) as well as a single administration (3 g/kg) suppressed the elevation of the plasma free radical metabolite levels after radiation (**Figure 19B**), suggesting that radiation-induced free radical productions may be effectively reduced.

5. Conclusions

Vitamin C is a strong antioxidative agent with potent radio-protective effects. In addition, vitamin C is a water-soluble vitamin and can be easily and safely excreted into the urine and can be excessively ingested. When bodies are exposed to radiation, the exposure evokes free radicals and ROS, which oxidize cell components, resulting in impairment of host cells. Antioxidants such as vitamin C protect hosts from radiation damage by scavenging such radical products. However, large doses of vitamin C are required to induce this radio-protective effect. We treated irradiated mice with p.o. administration with 150 or 250 mg/kg/day of vitamin C for

7–10 days [18, 19]. We also delivered i.p. administration with 3 g/kg of vitamin C to mice for postexposure treatment [20]. Because the daily requirement of doses of vitamin C is several milligrams per kilogram order in humans, several hundred to thousand times the doses of vitamin C were administered p.o. or i.p. to the mice in our studies.

A number of reports have described high-dose vitamin C therapy for the common cold, thermal injury, and malignant tumors, among other ailments [21, 25, 26]. In these reports, no severe side effects were observed by p.o. administration with high-dose vitamin C, although diarrhea or loose stool was occasionally noted. This is presumed to be osmotic diarrhea. In our unpublished data, when human volunteers took 5 g of vitamin C orally, approximately 7% of them had diarrhea or loose stool. The intravenous administration of high-dose vitamin C is widely used as a complementary and alternative medicine for various patients [27]. However, we should practice caution when administering gram-orders of vitamin C to patients [28]. Nephropathy due to oxalate, one of the main metabolites of vitamin C, has been reported in patients with renal impairment after the massive administration of vitamin C [29–31]. Patients with glucose-6-phosphate dehydrogenase deficiency also reportedly developed intravascular hemolysis after receiving massive administration of vitamin C [32, 33]. However, a recent clinical study has shown that the intravenous administration of 1.5 g/kg of vitamin C thrice weekly is safe and nontoxic in cancer patients, when patients with renal failure or glucose-6-deficiency were excluded [34].

We believe that vitamin C is a practical radioprotectant, as it is not expensive to synthesize and does not induce serious side effects, even if administered at a high dose. Indeed, several first responders dispatched to the Fukushima Dai-ichi Nuclear Power Plant Accident took vitamin C before their mission. Fortunately, no responders were exposed to high-dose radiation or developed acute radiation syndrome, but this prevented any evaluation of the effectiveness of vitamin C for preventing acute radiation syndrome; notably, though, no one reported any side effects.

If we use radiation or radioactive substance properly, it is an effective and a useful tool for us. However, fears still exist about misuse of this tool by someone with evil thought. In addition, we should always keep in mind that the human make mistakes without malice. We must prepare for radiological accidents without demanding any sacrifice by first responders.

Author details

Tetsuo Yamamoto[1] and Manabu Kinoshita[2*]

*Address all correspondence to: manabu@ndmc.ac.jp

1 Military Medicine Research Unit, Test and Evaluation Command, Ground Self-Defense Force, Setagaya, Tokyo, Japan

2 Department of Immunology and Microbiology, National Defense Medical College, Tokorozawa, Saitama, Japan

References

[1] Talbott EO, Youk AO, McHugh KP, Shire JD, Zhang A, Murphy BP, et al. Mortality among the residents of the three mile island accident area: 1979-1992. Environmental Health Perspectives. 2000;**108**(6):545-551

[2] Baranov A, Gale RP, Guskova A, Piatkin E, Selidovkin G, Muravyova L, et al. Bone marrow transplantation after the Chernobyl nuclear accident. The New England Journal of Medicine. 1999;**321**(4):205-212. DOI: 10.1056/NEJM/198907273210401

[3] Beresford NA, Fesenko S, Konoplev A, Skuterud L, Smith JT, Voigt G. Thirty years after the Chernobyl accident: What lessons have we learnt? Journal of Environmental Radioactivity. 2016;**157**:77-89. DOI: 10.1016/j.jenvrad/201602003

[4] Akashi M, Hirama T, Tonosaki S, Kuroiwa N, Nakagawa K, Tsuji H, et al. Initial symptoms of acute radiation syndrome in the JCO criticality accident in Tokai-mura. Journal of Radiation Research. 2001;**42**:S157-166

[5] Hirama T, Tanosaki S, Kandatsu S, Kuroiwa N, Kamada T, Tsuji H, et al. Initial medical management of patients severely irradiated in the Tokai-mura criticality accident. British Journal of Radiology. 2003;**76**(904):246-253. DOI: 10.1259/bjr/82373369

[6] Yasui S. Lessons learned: medical and health care management for emergency workers at the TEPCO Fukushima Daiichi APP accident. Journal of Occupational and Environmental Hygiene. 2014;**11**(5):49-58. DOI: 10.1080/15459624.2014.880445

[7] Scholes G. Radiation effects on DNA. The Silvanus Thompson Memorial Lecture. April 1982. British Journal of Radiology. 1982;**56**:221-231. DOI: 10.1259/0007-1285-56-664-221

[8] Prise FM, Davies S, Michael BD. A comparison of the chemical repair rates of free radical precursors of DNA damage and cell killing in Chinese hamster V79 cells. International Journal of Radiation Biology. 1992;**61**:721-728

[9] Straub JM, New J, Hamilton CD, Lominska C, Shnayder Y, Thomas SM. Radiation-induced fibrosis: mechanisms and implications for therapy. Journal of Cancer Research and Clinical Oncology. 2015;**141**(11):1985-1994. DOI: 10.1007/s00432-015-1974-6

[10] Ogawa Y. Paradigm shift in radiation biology/radiation oncology-exploitation of the "H_2O_2 effect" for radiotherapy using low-IET (linear energy transfer) radiation such as X-rays and high-energy electrons. Cancers. 2016;**8**(28):E28. DOI: 10.3390/cancers8030028

[11] Koyama S, Kodama S, Suzuki K, Matsumoto T, Miyazaki T, Watanabe M. Radiation-induced long-lived radicals which cause mutation and transformation. Mutation Research. 1998;**421**(1):45-54

[12] Pisani A, Riccio E, Andreucci M, Faga T, Ashour M, Di iNuzzi A, et al. Role of reactive oxygen species in pathogenesis of radiocontrast-induced nephropathy. BioMed Research International. 2013;**2013**:868321. DOI: 10.1155/2013/868321

[13] Guelman LR, Cabana JI, del Lujan Pagotto RM, Zieher LM. Ionizing radiation-induced damage on developing cerebellar granule cells cultures can be prevented by an early amifostine post-treatment. International Journal of Developmental Neuroscience. 2005;**23**(1):1-7. DOI: 10.1016/j.ijdevneu.2004.10.001

[14] Hensley ML, Schuchter LM, Lindley C, Meropol N, Cohen GI, Broder G, et al. American society of clinical oncology clinical practice guidelines for the use of chemotherapy and radiotherapy protectants. Journal of Clinical Oncology. 1999;**17**(10):3333-3355. DOI: 10.1200/JCO.1999.17.10.3333

[15] Jia D, Koonce NA, Griffin RJ, Jackson C, Corry PM. Prevention and mitigation of acute death of mice after abdominal irradiation by the antioxidant N-acetyl-cysteine (NAC). Radiation Research. 2010;**173**(5):579-589. DOI: 10.1667/RR2030.1

[16] Archambeau JO, Tovmasyan A, Pearlstein RD, Crapo JD, Batinic-Haberle I. Superoxide dismutase mimic, MnTE-2-PyP(5+) ameliorates acute and chronic proctitis following focal proton irradiation of the rat rectum. Redox Biology. 2013;**25**(1):599-607. DOI: 10.1016/j.redox.2013.10.002

[17] Khan S, Adhikari JS, Rizvi MA, Chaudhury NK. Radioprotective potential of melatonin against ^{60}Co gamma-ray-induced testicular injury in male C57BL/6 mice. Journal of Biomedical Science. 2015;**22**(61):1-15. DOI: 10.1186/s12929-015-0156-9

[18] Yamamoto T, Kinoshita M, Shinomiya N, Hiroi S, Sugasawa H, Matsushita Y, et al. Pretreatment with ascorbic acid prevents lethal gastrointestinal syndrome in mice receiving a massive amount of radiation. Journal of Radiation Research (Tokyo). 2010;**51**(2):145-156. DOI: 10.1269/jrr.09078

[19] Ito Y, Kinoshita M, Yamamoto T, Sato T, Obara T, Saitoh D, et al. A combination of pre- and post-exposure ascorbic acid rescues mice from radiation-induced lethal gastrointestinal damage. International Journal of Molecular Science. 2013;**14**:19618-19635. DOI: 10.3390/ijms141019618

[20] Sato T, Kinoshita M, Yamamoto T, Ito M, Nishida T, Takeuchi M, et al. Treatment of irradiated mice with high-dose ascorbic acid reduced lethality. PLoS One. 2015;**10**(2):e0117020. DOI: 10.137/journal.pone.0117020

[21] Cameron E, Pauling L. Supplemental ascorbate in the supportive treatment of cancer: Reevaluation of prolongation of survival times in terminal human cancer. Proceedings of the National Academy of the United States of America. 1978;**75**(9):4538-4542

[22] Carr AC, Vissers MCM. Synthetic or food-derived vitamin C—Are they equally bioavailable?. Nutrients. 2013;**5**:4284-4304. DOI: 10.3390/nu5114284

[23] Bolis G, Zanaboni F, Vanoli P, Russo A, Franchi M, Scarfone G, et al. The impact of whole-abdomen radiotherapy on survival in advanced ovarian cancer patients with minimal residual disease after chemotherapy. Gynecological Oncology. 1990;**39**(2):150-154

[24] MacVittie TJ. Therapy of radiation injury. Stem Cells. 1997;**15**(Suppl 2):263-268. DOI: 10.1002/stem.5530150735

[25] Cameron E, Pauling L. Supplemental ascorbate in the supportive treatment of cancer: Prolongation of survival times in human cancer. Proceedings of the National Academy of the United States of America. 1976;**73**(9):4538-4542

[26] Levine M, Conry-Cantilena C, Wang Y, Welch RW, Washko PW, Dhariwal KR, et al. Vitamin C pharmacokinetics in healthy volunteers: Evidence for a recommended dietary allowance. Proceedings of the National Academy of the United States of America. 1996;**93**(8):3704-3709

[27] Padayatty SJ, Sun AY, Chen Q, Espey MG, Drisko J, Levine M. Vitamin C: Intravenous use by complementary and alternative medicine practitioners and adverse effects. PLoS One. 2010;**5**(7):e11414. DOI: 10.137/journal.pone.0011414

[28] Stephanson CM, Levin RD, Spector T, Lis CG. Phase 1 clinical trial to evaluate the safety, tolerability, and pharmacokinetics of high-dose intravenous ascorbic acid in patients with advanced cancer. Cancer Chemotherapy and Pharmacology. 2013;**72**(1):139-146. DOI: 10.1007/s00280-013-2179-9

[29] McAllister CJ, Scowden EB, Dewberry FL, Richman A. Renal failure secondary to massive infusion of vitamin C. Journal of the American Medical Association. 1984;**252**(13):1684

[30] Lawton JM, Conway LT, Crosson JT, Smith CL, Abraham PA. Acute oxalate nephropathy after massive ascorbic acid administration. Archives of Internal Medicine. 1985;**145**(5):950-951

[31] Wong K, Thomson C, Bailey RR, Mcdiarmid S, Gardner J. Acute oxalate nephropathy after a massive intravenous dose of vitamin C. Australian and New Zealand Journal of Medicine. 1994;**24**(4):410-411

[32] Campbell GD Jr, Steinberg MH, Bower JD. Letter: Ascorbic acid-induced hemolysis in G-6-PD deficiency. Annals of Internal Medicine. 1975;**82**(6):810

[33] Rees DC, Kelsey H, Richards JD. Acute haemolysis induced by high dose ascorbic acid in glucose-6-phosphate dehydrogenase deficiency. British Medical Journal. 1993;**306**(6881):841-842

[34] Hoffer LJ, Levine M, Assouline S, Melnychuk D, Padayatty SJ, Rosadiuk K, et al. Phase 1 clinical trial of i.v. ascorbic acid in advanced malignancy. Annals of Oncology. 2008;**19**(11):1969-1974. DOI: 10.1093/annonc/mdn377

Section 3

Anticancer Effect of Vitamin C

Chapter 4

Vitamin C Against Cancer

Domenico Mastrangelo, Lauretta Massai,

Giuseppe Fioritoni and Francesco Lo Coco

Additional information is available at the end of the chapter

http://dx.doi.org/10.5772/intechopen.68746

Abstract

The selective anticancer properties of vitamin C are known since at least four decades. However, only recently *in vitro* studies have shown that vitamin C, in high enough concentrations, can efficiently and selectively kill a number of different human tumor cell lines, and these data have been confirmed in experimental animal tumor models. The first human clinical trials revealed that high doses of vitamin C administered by intravenous injection are not only very well tolerated but also substantially improve the quality of life of patients with clinically advanced cancer. However, the clinical evidence of the effectiveness of vitamin C in fighting off cancer is still controversial. The present chapter outlines the importance of vitamin C for a number of physiological functions, within the human body, and shows that there is a solid rationale for its use in the routine treatment of cancer, either alone or in combination with conventional treatment.

Keywords: vitamin C, sodium ascorbate, cancer, oxidative stress, free radicals, high dose intravenous ascorbate

1. Historical background

The anticancer effects of vitamin C (ascorbic acid) are known since 1969, when Benade et al. published a paper showing that the sodium salt of this nutrient (sodium ascorbate) is highly toxic or lethal to Ehrlich ascites carcinoma cells *in vitro* [1].

A few years before this discovery, the American biochemist, Irwin Stone, had already published some interesting reports on the genetic origin of scurvy, the lethal disease produced by severe deficiency of vitamin C. He had also coined the term "*hypoascorbemia*," to define the inability of humans and a few other species to synthesize vitamin C because of the lack of the enzyme L-gulonolactone oxidase (GLO), due to an *"inborn error of carbohydrate metabolism."*

After decades of research in this field, Stone became convinced that given the lack of GLO, and the low amount of vitamin C introduced with food, man easily undergoes a condition of *"chronic subclinical scurvy* (CSS)," and CSS is our most widespread disease. The long-term biochemical outcome of CSS, according to the scientist, sets the stage for the development of the serious medical problems of later life, including, among others, cardiovascular diseases (CVDs), collagen diseases, and cancer. Clinical tests reveal that mega levels of vitamin C are useful in the prevention and treatment of cancer and other diseases. Moreover, to correct CSS at least 10 g of vitamin C per day depending upon the incident stresses is required. Under stress, the daily requirement of vitamin C may be up to 200–300 g/day.

Stone noted that in the past decades, "micro" daily levels of vitamin C had wiped out acute frank scurvy, but did not prevent the epidemic incidence of CSS, the more insidious and more dangerous, relatively asymptomatic form of scurvy. The full correction of CSS is, therefore, the first step in any preventive medicine procedure. Even though the recommended dietary allowances (RDA) [2] prescribe daily amounts of vitamin C in the order of a few milligrams, these, according to Stone, will only prevent the appearance of the terminal symptoms of scurvy, but will not do much else.

On this ground, Stone concluded that "vitamin C" is not a real "vitamin," and proposed the use of the term "ascorbate," to better define this missing human liver metabolite [3, 4].

Therefore, according to the scientist, cancer, as well as almost any other known human diseases, depends on both the inability of humans to synthesize vitamin C and the insufficient amount of the nutrient normally assumed with food, leading to a deficiency (*"hypoascorbemia"*), which, in the long term, transforms into CSS, thereby predisposing to all kind of diseases.

The evidence that among the mammals producing their own vitamin C, an unstressed 70 kg goat is capable of producing 13 g of this liver metabolite, [5] and much more under stress had convinced Stone that vitamin C RDAs were largely underestimated. Therefore, he proposed the use of mega doses or doses ranging from 300 to several thousand times, the amount suggested by the RDAs of the nutrient to treat and prevent different diseases including cancer [6].

However, although formally proposed by Stone, the therapeutic use of mega doses of vitamin C was not really a novelty.

In 1949, Frederick Klenner had reported the successful treatment of 60 cases of bulbar poliomyelitis, with high doses of vitamin C, administered by mouth and, simultaneously, by intramuscular and intravenous injection, continuously, for 72–96 hours, until the complete remission of the symptoms [7]. Klenner proposed his anti-Polio, vitamin C–based treatment after reading a series of studies published by Jungeblut, between 1935 and 1937 [8], but he also treated with success, a number of other viral diseases, by using the same high dose vitamin C protocol.

The plea for a substantial revision of the dosage of vitamin C used in clinics had already come from the Nobel Prize Albert Szent-Györgyi, the discoverer of vitamin C, who, in the introduction to the Stone's book, "The healing factor," wrote, *"The medical profession itself took a very narrow and wrong view. Lack of ascorbic acid caused scurvy, so if there was no scurvy there was no lack of ascorbic acid. Nothing could be clearer than this. The only trouble was that scurvy is not a first symptom of lack but a final collapse, a premortal syndrome, and there is a very wide gap between scurvy and full health ... But nobody knows what full health is! ... Full health, in my opinion, is the condition in which we feel best and show the greatest resistance to disease. This leads us into statistics, which demand organization. However, there is another, more individual difficulty. If you do not have sufficient vitamins and get a cold, and as a sequence pneumonia, your diagnosis will not be "lack of ascorbic acid" but "pneumonia." So you are waylaid immediately"* [3].

Therefore, Szent-Györgyi had already warned the medical establishment about the need to radically review vitamin C RDAs (in the order of milligrams) that he considered sufficient to prevent scurvy, but largely insufficient to grant a condition of "full health."

The twofold Nobel Laureate, Linus Pauling later formalized this concept, in an article published in 1974 [9]. In this article, Pauling, after illustrating in detail the arguments in favor of the use of mega doses of vitamin C to treat a number of different diseases, suggests that the RDA of 45 mg/day, be renamed minimum dietary allowance, and the recommended dietary intake be introduced, ranging from 250 to 4000 mg/day for adults.

In the same year, Cameron and Campbell published an article concerning the treatment of 50 patients with advanced cancer, with 10 g of vitamin C administered by vein for the first few days, and then by mouth, for the rest of their lives. The results of this study indicated that high doses of vitamin C are useful as a routine supportive measure to reinforce standard treatment of earlier and more favorable cases [10].

Two years later, Cameron and Campbell demonstrated that the use of the protocol proposed by Cameron and Pauling significantly prolonged the survival and improved the quality of life of terminal cancer patients [11], but their work raised a number of criticisms, especially focused on the randomization procedure. To respond to the critics, the authors decided to undertake a second investigation, but this new study further confirmed that patients on mega doses of vitamin C lived, on average, 251 days longer than the untreated controls [12]. The same authors [13] and a group of Japanese clinicians [14] later confirmed the results formerly obtained by Cameron and Pauling.

In an attempt to either duplicate or refute the results reported by Cameron and Pauling, the Mayo Clinic initiated another investigation, which seemed to disprove the efficacy of the mega doses of vitamin C against cancer [15]. However, according to Pauling, the inclusion criteria used by the Mayo Clinic scientists were not conformed to the ones he had used. In fact, the Mayo Clinic study included patients previously treated with chemotherapy that compromises the immune response, while a functioning immune system is, according to Pauling, a fundamental prerequisite for an effective anticancer action of mega doses of vitamin C.

Therefore, based on this and other criticisms, the Mayo Clinic group undertook a second clinical investigation that substantially confirmed the results of the first one [16], and this study represented, for the scientific community, the definitive evidence of the inefficacy of mega doses of vitamin C against cancer.

As an undoubted evidence of the biases affecting this study, it will be worth mentioning that the second clinical trial performed at Mayo included only patients affected by colorectal cancer (CRC), that are clearly not representative of the entire complex and variegated range of cancer types affecting humans. However, to remain confined to just CRC as a paradigm of cancer, it will be worth mentioning the recent reports showing that vitamin C in high doses kills BRAF and KRAS mutants of CRC, which are resistant to the standard chemotherapeutic regimens, thus, substantially disproving the results of the second Mayo Clinic investigation [17–19].

2. Mechanistic explanation of the anticancer properties of vitamin C

Vitamin C is an essential nutrient with a number of beneficial functions, for the organism, since it

- helps the metabolism of tyrosine, folic acid, and tryptophan;
- increases the elimination of cholesterol;
- contributes to the synthesis of catecholamines;
- helps the body to absorb and breakdown histamine;
- enhances the absorption of nonheme iron;
- promotes the synthesis of collagen (its most widely known physiological function);
- neutralizes free radicals (it is a reducing agent, "scavenger" of free radicals, and a founder among the natural antioxidants);
- protects the DNA from damage due to free radicals and mutagens;
- reduces the risk of premature death;
- fights off widespread environmental pollutants;
- prevents the development of nitrosamines, and much more.

Vitamin C is ubiquitous, but humans, guinea pigs, some primates, a particular type of fruit-eating bat, the majority of fishes and birds do not produce it, and therefore they depend on diet for the assumption and use of this fundamental nutrient [20].

Regarding the anticancer properties of vitamin C, different authors have proposed various mechanistic explanations; among others:

2.1. The prooxidant pathway

As mentioned in the previous section, the hypothesis that vitamin C in high concentrations, administered by intravenous infusion, acts as a prooxidant, rather than antioxidant, leading to the formation of H_2O_2, with consequent oxidative damage to cancer cells, was formerly proposed in 1969 [1]. According to this hypothesis, vitamin C kills cancer cells through the intracellular generation of toxic hydrogen peroxide (H_2O_2) produced upon its oxidation by the cells themselves.

Although relevant for clinical cancer treatment, this discovery remained "hidden" for more than four decades, and the theory of the prooxidant activity of vitamin C in high concentration was proposed again 36 years later, though with no mention of the original work done in 1969 [21, 22].

The theory was further investigated, very recently, by Yun et al. [17–19] who defined the chain of events leading vitamin C to behave as a prodrug of H_2O_2 thereby killing selectively cancer cells, both *in vitro* and *in vivo*. Briefly, vitamin C in high doses administered by intravenous injection exerts its selective cytotoxic effect on cancer cells, because, after parenteral administration, it behaves as a peroxide delivery system for the generation of sustainable ascorbate radical and H_2O_2 in the extracellular space, with consequent oxidative damage to cancer cells (**Figure 1**). The selectivity of the cytotoxic effect of parenteral vitamin C depends on the fact that cancer cells, compared to their normal counterpart, show a reduced level of antioxidant enzymes such as catalase, superoxide dismutase, and glutathione peroxidase. The reduced level of antioxidant enzymes leads to cellular damage by accumulation of H_2O_2, with consequent intracellular redox imbalance and oxidative damage to different cellular structures.

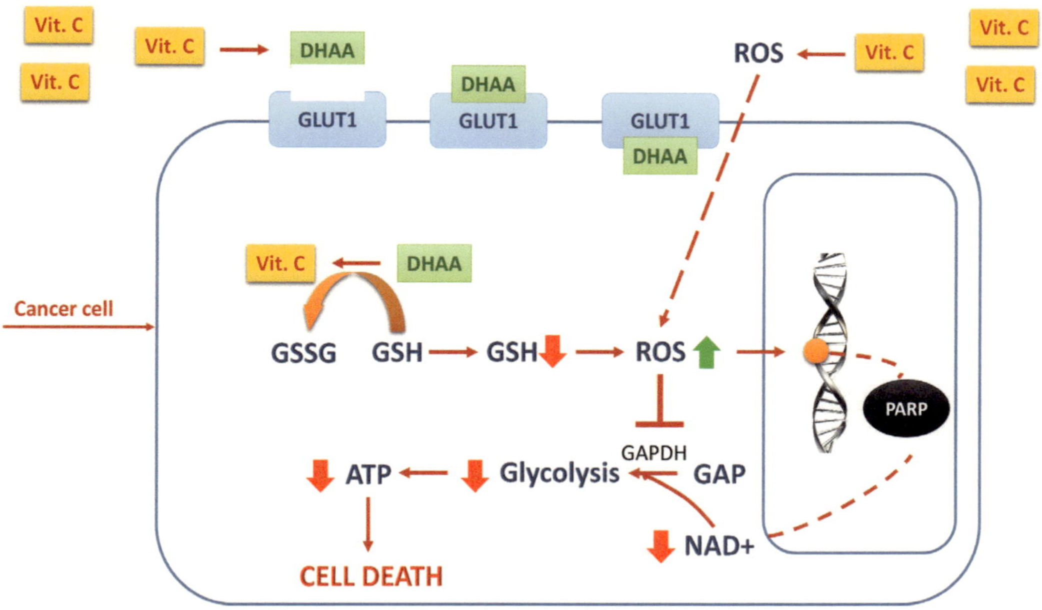

Figure 1. Prooxidant effect of vitamin C (see text). Legend: Vit. C = vitamin C, ATP = adenosin triphosphate, DHAA = dehydroascorbic acid, GSH = glutathione, GSSG = glutathione disulfide, GAP = glyceraldehyde 3-phosphate, GAPDH = glyceraldehyde 3-phosphate dehydrogenase, GLUT = glucose transporter, NAD = nicotinamide adenine dinucleotide, ROS = reactive oxygen species, PARP = poly ADP-ribose polymerase.

Yun et al. [17–19] have recently showed that the death of KRAS and BRAF cell mutants of CRC is imputable to the oxidized form of vitamin C: dehydroascorbic acid (DHAA). DHAA competes with glucose, for intracellular uptake by glucose transporters (GLUTs), mainly one and four subtype receptors. Interestingly, both KRAS and BRAF activating mutations are responsible of the upregulation of GLUT1 expression in different types of cancer, including CRC, although the upregulation of GLUT1 expression is not always associated with increased sensitivity of tumor cell lines to the cytotoxic effects of DHAA.

Investigation into the metabolic makeup of KRAS and BRAF CRC-derived cell lines shows that there is an accumulation of glycolytic intermediates upstream glyceraldehyde-3-phosphate-dehydrogenase (GAPDH) and a contemporary depletion of the metabolites downstream GAPDH, indicating an inhibition or severe reduction of its enzymatic activity, which appears to be the key of the cytotoxic effect of DHAA.

In summary, the data reported by Yun et al. on the effect of DHAA on CRC cell lines indicate that in glycolysis-addicted KRAS and BRAF mutated cell lines, high amounts of DHAA enter the cancer cells, thanks to the overexpressed GLUT-1 receptors. DHAA is then reduced again to vitamin C inside the cells. The reduction of DHAA to vitamin C scavenges glutathione (GSH), thus inducing redox imbalance and oxidative stress. Oxidative stress, in turn, leads to inactivation of GAPDH, inhibition of glycolysis, and energetic crisis, which leads to cancer cell death.

According to this mechanistic explanation, vitamin C, functioning as a prooxidant, would induce an increase in the intracellular reactive oxygen species (ROS), which leads to increased DNA damage, with consequent activation of poly ADP-ribose polymerase (PARP), an enzyme necessary to repair damaged DNA. PARP activation would in turn consume NAD+, with NAD+ depletion and consequent ADP depletion, leading to energetic crisis and death of cancer cells [23].

The theory according to which vitamin C in high doses would act as a prodrug of H_2O_2, beyond being criticized by different authors, seems somewhat controversial and overlooks a few important aspects as follows:

- Although reported in 2005 by Chen et al. [21, 22], it is not new, since, as we have seen, it had been already proposed by Benade et al. in 1969 [1]. Interestingly, while according to Benade, H_2O_2 forms inside the cell, starting from vitamin C, Chen et al. conclude that H_2O_2 is formed outside the cell, staring from DHAA. Our experience with high concentrations of vitamin C to treat, *in vitro*, retinoblastoma (Y79) (**Figures 2** and **3**) [24], uveal melanoma (C918, OCM1) [25], human promyelocytic leukemia (HL60) [26], and different human myeloid leukemia (HL60, U937, K562, NB4, NB4-R1, and NB4/As) cell lines [27] indicates that H_2O_2 forms inside the cells, rather than outside. In fact, in our experiments, the cytotoxic effect of vitamin C on cancer cells in culture persists for hours/days after the removal of vitamin C from the culture medium.
- H_2O_2 is a metabolite normally produced by the cells of the body and usually overproduced by cancer cells. Therefore, H_2O_2 itself could be an optimal substitute for vitamin C, as an anticancer compound. In this regard, it could be useful remark that Reginald Holman, in 1957, published a paper in "Nature" showing that rat implanted with Walker 256 adenocarcinoma

http://dx.doi.org/10.5772/intechopen.68746

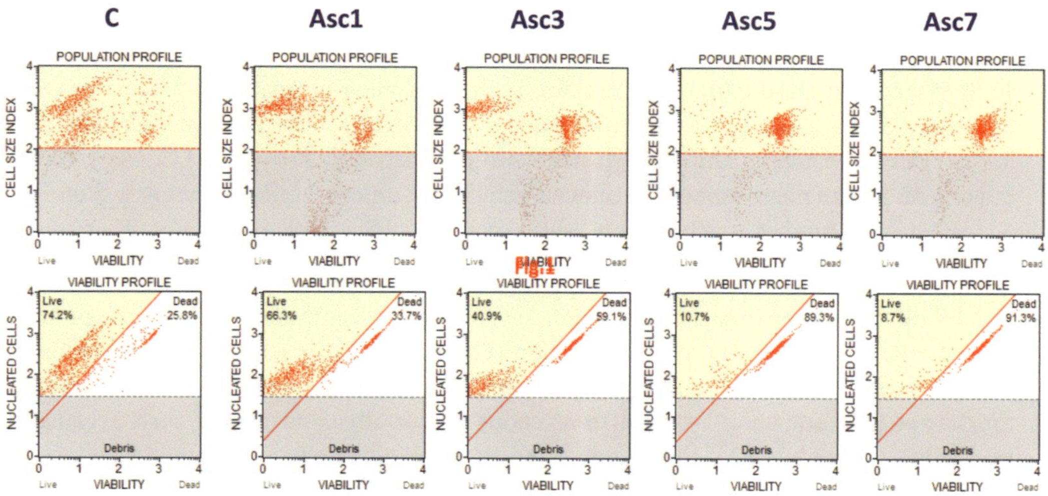

Figure 2. Flow cytometric analysis of Y79 human retinoblastoma cell line viability, after treatment with increasing concentrations of vitamin C (Asc) *in vitro*. C = control sample, Asc1 = vitamin C 1 mM, Asc3 = vitamin C 3 mM, Asc5 = vitamin C 5 mM, Asc7 = vitamin C 7 mM. Starting from a viability of about 74% (control sample), the percentage of viable cells after 1 hour of treatment with vitamin C and 18–24 hours of incubation are about 66% at 1 mM, 41% at 3 mM, 11% at 5 mM, and 9% at 7 mM of vitamin C.

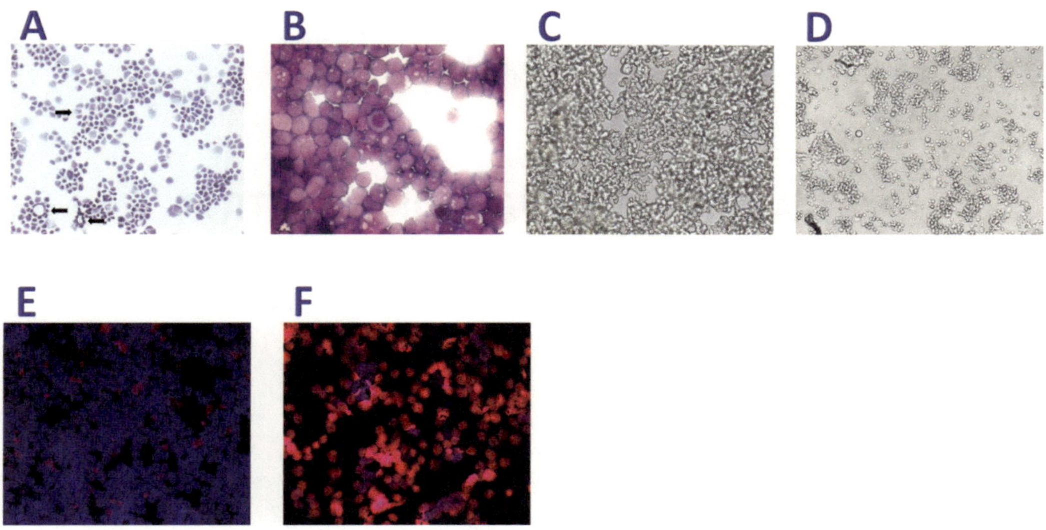

Figure 3. Microphotographs of Y79 human retinoblastoma cell lines: A: hematoxylin/eosin staining of cultured Y79. Interestingly, cells in culture tend to form the typical Flexner-Wintersteiner «rosettes» (black arrows) commonly seen in pathology specimens (original magnification: 200×). B: May-Grunwald Giemsa staining showing morphological details: large cells with loose chromatin and highly basophilic cytoplasm (original magnification: 400×). C: Contrast phase microphotographs of Y79 human retinoblastoma cell lines in culture (control sample): cells tend to form clusters in culture (original magnification: 200×). D: Contrast phase microphotograph of Y79 human retinoblastoma cell lines after treatment with vitamin C 7 mM. At a concentration of 7 mM, vitamin C destroys the vast majority of cells in culture. A few residual cells appear swollen and necrotic (original magnification: 200×). E and F: Hoechst/PI staining of Y79 human retinoblastoma cell line in the culture. With Hoechst/PI, live cells not distinguishable in the black/withe caption. In the control sample (E), there is a clear prevalence of cells stained by the Hoechst dye (alive) while in the sample treated with 7 mM of vitamin C (F), the prevalent color is given by PI (dead cells)(original magnification: 200×).

and treated by simply replacing their drinking water with 0.45% hydrogen peroxide showed a rate of cure of 50–60% [28]. The time for complete disappearance of the tumor varied from 15 to 50 days depending on the tumor size at the beginning of treatment. Holman's work was based on the assumption (later confirmed by studies on vitamin C) that malignant cells are deficient in catalase, and as such unable to detoxify high fluxes of H_2O_2. As a consequence, an *in vivo* measurement of catalase activity in tumors would represent a useful diagnostic tool to predict which cancers will respond to pharmacological doses of vitamin C therapy (or H_2O_2) [29].

- According to the prooxidant theory, vitamin C in high concentrations induces the production of H_2O_2 through a Fenton-like reaction (**Figure 4**). This reaction is the oxidation of organic substrates by iron and hydrogen peroxide. Trivalent iron ($Fe3^{+}$) is fundamental for the reaction, but, since Fenton-like reactions are usually controlled, *in vivo*, because of iron sequestration by metal-binding proteins, the prooxidant effect of vitamin C, *in vivo*, is considered scarcely significant by different authors [30, 31], and other mechanisms should be hypothesized.
- Cancer cells produce high amounts of H_2O_2, and high levels of this metabolite have been associated with key features in cancer, such as DNA alterations, cell proliferation, apoptosis resistance, metastatic spread, angiogenesis, and hypoxia-inducible factor 1 (HIF-1) activation. On the contrary, decreasing the cellular levels of H_2O_2 may reverse the malignant phenotype. Therefore, H_2O_2 can be either proapoptotic or antiapoptotic, either carcinogenic or anticarcinogenic, depending on its concentration and localization within the cell [32, 33];

VITAMIN C AND FENTON REACTION

1. $Fe^{3+} + AscH_2 \longrightarrow Fe^{2+} + Asc^{-\cdot} + 2H^+$

2. $Fe^{2+} + O_2 \longrightarrow Fe^{3+} + O_2^{-\cdot}$

3. $2O_2^{-\cdot} + 2H^+ \longrightarrow H_2O_2 + O_2$

4. $H_2O_2 + Fe^{2+} \longrightarrow Fe^{3+} + OH^{\circ} + OH^-$

Figure 4. Fenton reaction mediated by vitamin C. (1) Vitamin C (ascorbic acid, AscH2) reduces ferric ions (Fe3+) to ferrous ions (Fe2+). (2) Ferrous ion reacts with oxygen to produce superoxide. (3) Dismutation of superoxide leads to hydrogen peroxide. (4) Hydrogen peroxide reacts with ferrous ions to form hydroxyl radicals. Legend: Asc = ascorbic acid, vitamin C, Fe = iron, O = oxygen, H_2O_2 = hydrogen peroxide, OH = hydroxyl radical.

http://dx.doi.org/10.5772/intechopen.68746

- According to the "Fenton chemistry," invoked to explain the selective cytotoxic effect of vitamin C against cancer cells, trivalent iron (Fe^{3+}) is necessary for the formation of H_2O_2, starting from vitamin C. However, some literature data seem to demonstrate that the exact opposite is true. In particular, Mojic et al. using two prostate cancer cell lines (LNCaP and PC-3) showed that iron at physiological concentrations in the cell culture medium and human plasma abrogates the anticancer/cytotoxic effects of vitamin C. According to these authors, iron at physiological concentrations promotes both production and decomposition of H_2O_2, the latter being mediated by a Fenton reaction, which prevents the accumulation of H_2O_2, thus abolishing the cytotoxic effect of vitamin C. Therefore, as the authors conclude, the *in vitro* investigations on the anticancer properties of vitamin C may have been overestimated because all suffered the bias of a low amount of Fe^{3+} in the culture medium, if compared to normal plasma and body fluids. To repeat *in vivo* the results obtained *in vitro*, the authors suggest that the simultaneous administration of vitamin C and chelating agents remove iron [34].
- Vitamin C (ascorbate) readily undergoes pH-dependent autoxidation producing hydrogen peroxide, and catalytic metals accelerate the oxidation process. This means that catalytic iron is not strictly necessary for the production of H_2O_2, starting from vitamin C, and therefore, the Fenton reaction may not be essential for this purpose. The autoxidation, i.e., oxidation in the absence of catalytic metals, occurs via the ascorbate dianion (Asc^{2-}). In particular, that at pH 7.0, 99.9% of ascorbate (vitamin C) is in the form of monoanion ($AscH^-$). $Asc2^-$ increases by a factor 10, with one unit increase in the pH. Therefore, while the production of H_2O_2 may be scarcely relevant in the absence of catalytic iron (as in the "Fenton chemistry"), it may become considerable when the concentration of ascorbate is in the order of the millimoles, as in the case of the use of vitamin C as an anticancer compound. To give an example, an aqueous solution containing 20 mM of vitamin C in the form of sodium ascorbate (the sodium salt of the ascorbic acid) will contain 1 μM of Asc^{2-} which, in turn, will result in a flux of H_2O_2 on the order of 10 nM/s in a typical cell culture experiment [35].
- In most laboratory settings, autoxidation of vitamin C is due to adventitious catalytic metals, as part of the buffers used or contaminating of lab equipment. It is not by chance that the methods underpinning the "Fenton Paranoia" are *in vitro* methods using either isolated tissue cultures or d samples exposed to the air. Indeed, as reported by some authors: "... *unless extreme care is taken, every time vitamin C is added to blood samples (and urine samples) outside the body there are a whole host of oxidative products and markers produced* ..." [36].
- Both the antioxidant and prooxidant activities of vitamin C in high doses may not necessarily be mutually exclusive. Studies on chelation therapy have shown that 5 g of the sodium salt of vitamin C added to the ethylenediaminetetraacetic acid (EDTA) chelation cocktail results in acute oxidative stress, but this effect is transitory, and after multiple sessions of EDTA-based chelation treatment, a prolonged, protective, antioxidant effect of the treatment becomes evident [37]. These data confirm the evidence reported by Mojic et al.[34] regarding the inhibitory effect of iron on the prooxidant activity of vitamin C, and also the idea, formerly illustrated by Klenner, according to which vitamin C in high concentration may act as a "flash oxidizer" [38].

In summary, increasing the concentration of vitamin C with mega doses of the nutrient injected in vein may lead to a substantial increase in the spontaneous generation of H_2O_2, with toxic consequences to cancer cells, according to the chain of biochemical reactions more recently described by Yun et al., and briefly summarized herein. This does not mean that the administration of vitamin C in high doses by intravenous injection abrogates its antioxidant properties, but most probably that both prooxidant and antioxidant effects coexist. It rather implies that both antioxidant and prooxidant properties are simultaneously present in the molecule, the latter resulting more pronounced when the concentration of the nutrient reaches values in the order of millimoles (20 mM).

2.2. The antioxidant pathway

"Collagen" is a collective name to designate group of fibrous proteins that occur in vertebrates as the chief constituent of connective tissue fibrils and in bones. Many of the symptoms of scurvy (the syndrome of acute deficiency of vitamin C) depend on the defective production of collagen. Since the beginning of the history of this nutrient, scientists know that vitamin C is essential for the synthesis of collagen. Indeed, vitamin C is a cofactor of collagen prolyl-4-hydroxylase (C-P4H), the enzyme responsible for the formation of hydroxyproline, the essential component of collagen. Under conditions of vitamin C deficiency, C-P4H loses its activity, and the organism does not form collagen properly, with consequent connective tissue deterioration, as it happens in scurvy (**Figure 4**).

Prolyl-hydroxylases are an entire family of enzymes, also known as 2-oxoglutarate-dependent dioxygenases (2-OGDDs) with a wide range of biological functions [39], and members of this family include HIF-hydroxylases [40]. These vitamin C–dependent enzymes are of extreme importance in tumor biology since hypoxia and induction of HIFs are a hallmark of many tumors [41].

HIF is a heterodimeric transcription factor discovered in 1991 [42]. In normal oxygen pressure conditions ("normoxia"), the HIF-1α unit is downregulated by vitamin C–dependent hydroxylases, while in hypoxic conditions, such as those so frequently found in solid tumors, there is a repression of the hydroxylation of HIF-1α, with consequent increase of HIF-dependent gene transcription, neoangiogenesis, and tumor growth and progression. The important consequence of the central role of vitamin C in the synthesis of HIF-1α hydroxylases is that low levels of vitamin C promote tumor growth and progression (as already hypothesized by Stone, Szent-Györgyi, and Pauling), by reducing HIF-1α hydroxylation, thereby stabilizing HIF-1α [43], and high levels of HIF render cancer cells more sensitive to vitamin C–induced toxicity [44].

Recently, Kuiper et al. have confirmed the above data by finding an inverse relationship between intratumor levels of vitamin C and HIF activity in both endometrial cancer and CRC [45–47].

For a better understanding of the centrality of the relationship between hypoxia and HIF in tumor biology, we must consider that cancer hypoxia (a very common feature in cancer) is associated with HIF activity that mediates angiogenesis, epithelial-mesenchymal transition

(EMT), stem cell maintenance, invasion, metastasis, and resistance to radiation therapy and chemotherapy [48]. Therefore, attempts to downregulate HIF synthesis and activity may represent a step forward in the search of an effective and selective anticancer drug [49, 50].

Is vitamin C such a molecule?

The current evidence shows that vitamin C has a close relationship with the function of HIF, and therefore, being a natural compound, it is the best-suited, natural molecule for the purposes of inhibiting cancer growth through HIF-mediated mechanisms.

Tian et al. note that the overexpression of HIF greatly enhances vitamin C–induced toxicity on cancer cells, since HIF increases the intracellular uptake of oxidized vitamin C through its transcriptional target GLUT1, synergizing with the uptake of its reduced form through sodium-dependent vitamin C transporters (SVCTs) [44].

Other authors, working with human leukemia cell lines, showed that vitamin C inhibits the growth of human leukemic cells not only through the generation of H_2O_2 but also and more importantly through the downregulation of HIF-1α transcription [51].

Further important insights into the role of HIF have come from studies of three tumorigenic models *in vivo*, showing that the antitumorigenic effects of antioxidants such as N-acetylcysteine (NAC) and vitamin C are not due to their ability to squelch DNA damage and genomic instability mediated by ROS but due to their capacity to downregulate HIF levels [52].

These results are of extreme interest for the following reasons:

- Whether we consider its prooxidant activity, leading to cancer cell damage through the generation of H_2O_2 or its antioxidant (more typical) activity, leading to the enzymatic breakdown, and nonenzymatic downregulation of the HIF, vitamin C in high doses always shows the characteristics of a simple, natural, and effective anticancer molecule;
- Antioxidants (all of them!) have anticancer effects;
- Vitamin C and other antioxidants may have a role as adjuvant therapy to prevent progression or recurrence of HIF-dependent tumors;
- Vitamin C may show anticancer properties even when administered by mouth, with caution about the dose, which must be sufficiently high.

Recent investigations have shown that ascorbate therapy has a significant effect on the expression of several genes relevant to the development or inhibition of cancer. In particular, the reduced expression of tumor-promoting genes, such as HIF, and the increased expression of tumor-suppressor genes, such as p53, support the notion that vitamin C can act as a potential agent for the suppression of tumor development [53].

2.3. The epigenetic pathway

As we have seen, vitamin C has a number of beneficial biological functions, many of which are related to its action as an electron donor for adjusting the redox state of iron-containing enzymes. Recent studies have implicated Fe^{2+}-dependent oxidative modification activities in normal tissue homeostasis and experimentally induced reprogramming. The loss of these activities is associated with epigenetic defects and compromised cell differentiation or developmental potential (**Figure 5**).

One of the most important properties of Fe^{2+}-2OG-dependent dioxygenases is their susceptibility to environmental factors (as we have seen in the case of HIF-hydroxylases). Vitamin C is necessary to maintain the function of these enzymes, and different oncometabolites can inhibit their activity.

Recent studies show that vitamin C may enhance reprogramming of pluripotent stem cells, and the available data suggest a strong link between vitamin C, dioxygenase function, and stem cell differentiation, that is of great relevance for human disease [54]. In fact, vitamin C (in the form of sodium ascorbate), as a member of Fe^{2+}- and 2OG-dependent dioxygenases, plays a critical role in the demethylation of DNA and histone, as a cofactor for a group of enzymes termed methylcytosine dioxygenase ten-eleven translocation (TET, including TET1, TET2, and TET3) and some Jumonji-C (JmjC) domain-containing histone demethylases [55, 56].

(**Note 1**: TET1 catalyzes the conversion of the modified DNA base 5-methylcytosine (5-mC) to 5-hydroxymethylcytosine (5-hmC). TET1 produces 5-hmC by oxidation of 5-mC in an iron and alpha-ketoglutarate-dependent manner [8]. The conversion of 5-mC to 5-hmC has been proposed as the initial step of active DNA demethylation in mammals.)

(**Note 2**: TETs belong to the Fe^{2+}- and 2OG-dependent dioxygenase superfamily.)

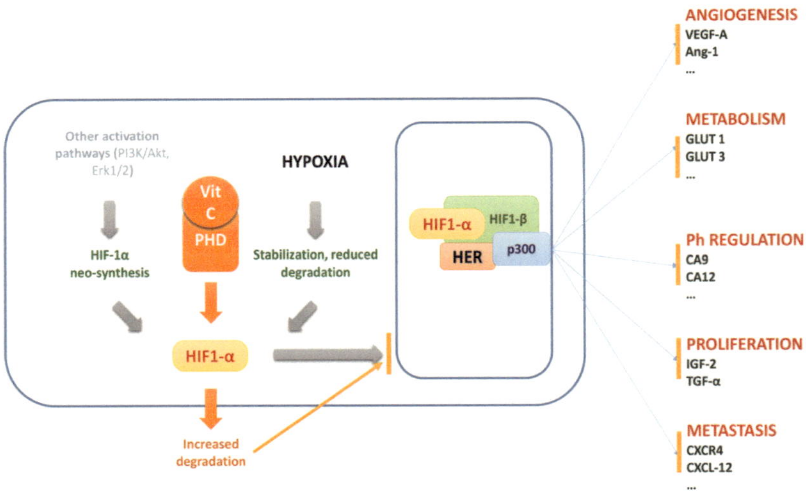

Figure 5. Hypoxia stabilizes HIF-1α as the rate of prolyl-hydroxylases (PHD)-mediated hydroxylation of a subunit is limited. HIF-1α accumulates in the cytoplasm and passes into the nucleus, where it binds to HIF1-β. Cofactor p300 allows the binding to hypoxia response elements (HREs) of target genes involved in angiogenesis, metabolism, pH regulation, proliferation, and metastasis. Vitamin C, as a cofactor of PHD, determines increased degradation of HIF1-α and, as a consequence, inhibits the expression of genes activated by HIF1-α.

Given its role as an essential cofactor of enzymes involved in epigenetic gene regulation/expression, vitamin C can be involved in embryonic development, postnatal development, aging, cancer, and other diseases [57].

In summary, the epigenetic gene regulation functions of vitamin C encompass:

- The regulation of DNA demethylation as an essential cofactor for TET dioxygenases;
- The regulation of histone demethylation as an essential cofactor for JmjC domain-containing histone demethylases;
- The role of interface between the genome and environment;
- The critical role in maintaining the epigenome, especially at early embryonic stages;
- The contribution to different disease in the case of deficiency, which causes reduction of the catalytic activity of TET dioxygenases and JmjC domain-containing histone demethylases [58].

According to some authors, vitamin C causes widespread, consistent, and remarkably specific DNA demethylation of 1847 genes in human embryonic stem cells (hESCs), including important stem cell genes, with a clear bias toward demethylation at CpG island boundaries [59].

Other epigenetically relevant effects of vitamin C include:

- The development of dopamine neuron differentiation in fetal midbrain [60], the induction of pluripotent state in mouse embryonic stem cells [61, 62];
- The enhancement of the demethylating activity of 5-azacytidine, and induction of cytotoxicity [63, 64];
- The inhibition of the malignant phenotype on melanoma cells *in vitro*, by partially reestablishing the global content of 5-hydroxymethylcytosine (5-hmC) and the consequent alteration in the transcriptome [65];
- The upregulation of several microRNA (miRNA) involved in tumor suppression and drug resistance, the most prominent of which correlates with increased overall survival of breast cancer or nasopharyngeal carcinoma [66];
- The inhibition of the proliferation, migration, and epithelial-mesenchymal-transition (EMT) of lens epithelial cells by destabilizing HIF-1α [67].

2.4. The immunologic pathway

Vitamin C concentrations in the plasma and leukocytes rapidly decline during infections and stress, and vitamin C supplementation improves a number of different immunologic functions, including, among others: antimicrobial and natural killer (NK) activities, lymphocyte proliferation, chemotaxis, and delayed-type hypersensitivity. Furthermore, by maintaining the cellular redox integrity, vitamin C protects the cells of the immune system against ROS [68, 69]. Immunocompetent cells such as lymphocytes, neutrophils, and monocytes have vitamin C levels 10—100-fold higher than the plasma, and accumulate it against a concentration gradient (**Figure 6**) [70].

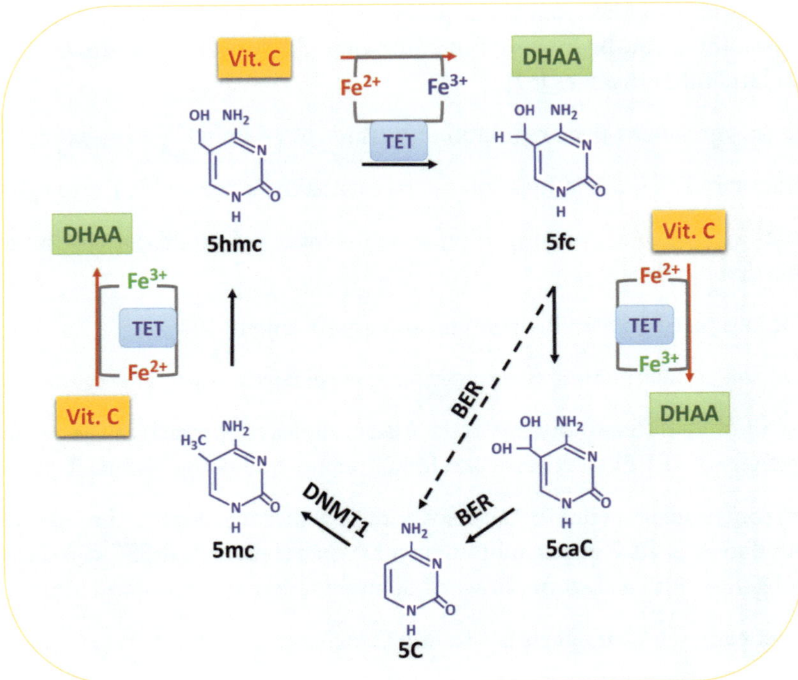

Figure 6. As a cofactor for TET dioxygenases, vitamin C participates in the conversion of 5-mC to 5-hmC, and further 5fC and 5caC, thus modulating DNA demethylation. Legend to figure: 5C = cytosine, 5-mC = 5-methylcytosine, 5-hmC = 5-hydroxymethylcytosine, 5fC = 5-formylcytosine, 5caC = 5-carboxylcytosine, BER = base excision repair, DNMT1 = DNA methyltransferase-1, DHAA = dehydroascorbic acid, Vit. C = vitamin C.

A decrease in the intracellular content of vitamin C may result in locally increased apoptosis of immune cells and immunosuppression [71].

Vitamin C is essential for immunoglobulin synthesis [72] and active phagocytosis [73] enhances the production of interferon [74] and suppresses the production of interleukin-18 (IL-18), a key regulator in malignant skin tumors, including melanomas, squamous cells carcinomas, and a number of other tumors [75].

The concept that the immune system can help fighting cancer has deep roots. In 1909, the German scientist Paul Ehrlich proposed that the incidence of cancer would be much higher, were it not for the action of our immune system in recognizing and eliminating tumor cells [76].

Half a century later, two scientists, Lewis Thomas and Frank Macfarlane Burnet, took Paul Ehrlich's original idea a step further and proposed the model of "immune surveillance," according to which the cells of the immune system actively patrol the body looking for cancerous cells and eliminate them as they arise. This idea became a grounding principle of the new field of cancer immunology that took shape beginning in the 1950s [77].

More recently, the relationship between cancer and the host immune system has become clear with the introduction of the concept of "immune editing," a three-phase process leading cancer cells to escape the control of the immune system, with consequent cancer progression [78].

Therefore, since no doubts exist, nowadays, about the role of the immune system in controlling cancer development, progression and spread, natural substances such as vitamin C, whose action spreads over a wide range of effectors of both the innate and adaptive immunity, represent a new promising therapeutic tool against cancer.

Much of the "booster" action of vitamin C on the immune system depends on its antioxidant nature. Indeed, vitamin C downregulates ROS-dependent expression of pro-inflammatory interleukin genes, via inhibition of transcription of NF-κB (nuclear factor kappa-light chain-enhancer of activated B cells), which, in turn, regulates the expression of pro-inflammatory cytokines, such as IL-1 and tumor necrosis factor-alpha (TNFα) [79]. Recent evidence shows that vitamin C enhances antioxidant defenses of T-cells [80] and increases T-cell responsiveness to antigens, thus suggesting that it has a definite role in regulating immune function [81].

Confirm of the role of vitamin C in regulating the immune system function come from studies showing that increased vitamin C concentrations inhibit antigen-induced, withdrawal-induced, steroid-induced, and spontaneous T-cell apoptosis [82], fas-induced apoptosis of monocytes [83], and increased cytotoxic activity of natural killer cells in humans [84]. All these data indicate that it can modulate the immune system by inhibiting T-cell apoptosis signaling pathways [85]. Similar results apply to monocyte-derived dendritic cells (DCs) [86].

As mentioned in the "historical background," more than four decades ago, Pauling attributed to vitamin C the role of a "booster" of the immune system. This was his main argument against the negative results reported by the Mayo Clinic scientists, since they should not have included, in their clinical trials, cancer patients previously treated with chemotherapy, because chemotherapy destroys immunocompetent cells and severely impairs the individual immune response to cancer.

The Mayo Clinic scientists apparently did not understand this argument and, in reply to the legendary scientist wrote "*… Any contention that previous chemotherapy prevented our patients from achieving the extraordinary survival increase claimed by Drs. Cameron and Pauling must be considered highly speculative at best. Our patients were entered into the study only when they were well past any acute immune-suppressive effects of previous therapy*" [87].

The evidence, today, indicates that a healthy immune system is necessary to control the malignant disease and that immune suppression associated with cancer and cancer chemotherapy contributes to its progression [88]. Therefore, in agreement with Pauling and the current evidence, it is likely to assume that since cancer chemotherapy produces long-lasting deleterious effects on the immune system [89, 90], patients previously treated with cytotoxic chemotherapy should not expect positive outcomes from the treatment with vitamin C as a "booster" of the immune system.

2.5. The collagen pathway

William McCormick first proposed, about 6 decades ago, the hypothesis that cancer could be due to a defect in the metabolism of collagen [91]. Given the fundamental role of vitamin C, in the synthesis of collagen [92], the scientist concluded that a deficient intake of vitamin C (as in full blown or chronic subclinical scurvy) could determine cancer by disrupting the synthesis of collagen, thus creating a local permissive environment for cancer to grow and spread to other organs and tissues (**Figure 7**) [93].

As we have seen in the previous sections, vitamin C is required for collagen synthesis by acting as a cofactor for nonheme iron α-ketoglutarate-dependent dioxygenases such as prolyl-4-hydroxylase. It stimulates all types of collagen synthesis by donating electrons required for hydroxylation reactions of proline and lysine in procollagen, normally performed by specific hydroxylating enzymes.

In recent years, the role of basement membrane (BM) in the dynamic regulation of cell behavior and cell-signaling pathways has clearly emerged. The BM defines the tumor microenvironment and provides significant host-derived regulatory signals during progression of tumor growth and metastasis. The major component of basement membrane is type IV collagen, and recent studies indicate that in cancer progression there is a disruption of the normal assembly and organization of the basement membrane [94].

These studies support the notion that the loss of collagen type IV chains may provide a permissive microenvironment for cancer invasiveness, as hypothesized by McCormick, more than 5 decades ago [95].

Metastatic spread is a crucial event in the evolution of cancer, being responsible of the majority of cancer-related deaths. The steps leading to metastatic spread of cancer encompass the detachment of cancer cells from the primary tumor, the disruption of the BM, the invasion of the surrounding stroma, cancer cell entry into and transport through the vascular or lymphatic

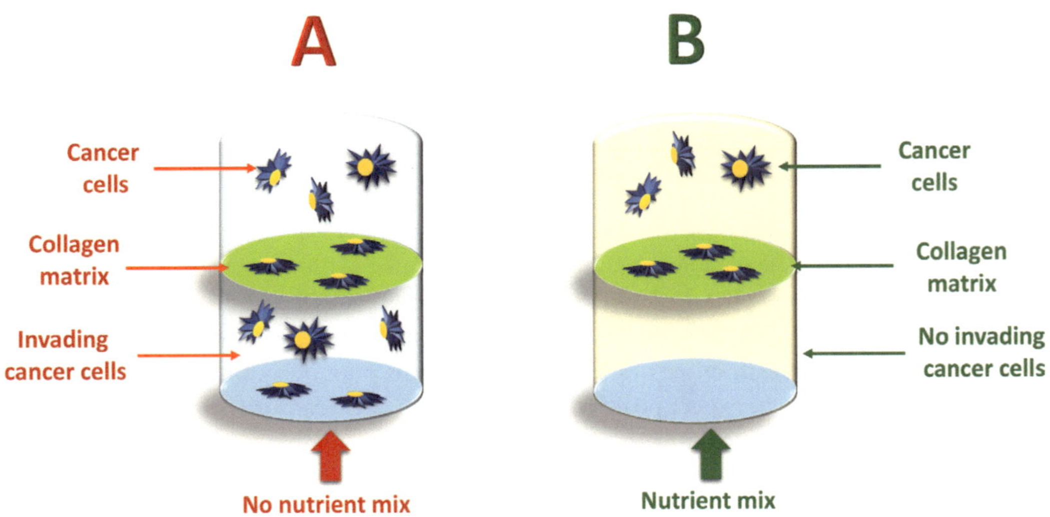

Figure 7. The "nutrient mix" used is a mixture of vitamin C, L-lysine, L-proline, and epigallocatechin gallate (EGCG). Cancer cells can destroy enzymatically the collagen matrix (A) between the two chambers of the vial, and spread over the second chamber. When the nutrient mix is added (B), cancer cells are no longer able to destroy the collagen matrix and therefore they remain confined to the superior chamber (see Refs. [99, 100]).

system to distal sites (liver, lungs, brain, etc.), and extravasation, tumor cell proliferation and angiogenesis at distal sites [96].

A critical event in tumor cell invasion is the degradation of the extracellular matrix (ECM), a complex network of extracellular macromolecules such as collagen, proteoglycans, fibronectin, laminin, and other glycoproteins that act as a barrier to the spread of cancer cells to distal sites by restricting tumor growth and invasion.

Matrix metalloproteinases (MMPs) are calcium-dependent endopeptidases, which require coordination of a zinc ion to mediate catalysis. As implied by their name, MMPs operate on a variety of substrates belonging to the ECM [97], and owing essentially to their vast diversity, MMPs participate in nearly every biological process, involving the remodeling of the ECM, from implantation of an embryo into the uterine wall to tissue necrosis.

A major structural protein for ECM and basement membrane is type IV collagen. Therefore, type IV collagenases MMP-2 (72-kd gelatinase A) and MMP-9 (92-kd gelatinase B), usually overexpressed in malignancy, are the focus of research in this field.

All types of cancer cells form tumors and spread in the body by degrading the ECM by means of various matrix MMPs. The activity of these enzymes correlates with the aggressiveness of tumor growth and invasiveness of the cancer.

In 1992, Pauling and Rath hypothesized that natural compounds such as lysine and vitamin C could inhibit ECM proteolysis and, as such, had the potential to modulate cancer growth and spread [98]. These nutrients would exert their anticancer activity by both inhibiting MMPs and strengthening the connective tissue surrounding the tumor.

Several lines of evidence support an indispensable role for vitamin C in maintaining good-quality collagen. Vitamin C assists the posttranslational modification of collagen by reducing iron in the participating enzymes, lysyl-hydroxylase, and prolyl-hydroxylase. Experimental *in vitro* data show incubation of cancer cells with a nutrient mixture containing vitamin C, L-lysine, L-proline, and epigallocatechin gallate (EGCG), they are no longer able to invade the collagen matrix [99, 100] and spread at distant sites (**Figure 7**).

Although the mechanisms through which the nutrient mixture used in these experiments inhibits MMPs deserves further investigation, it is quite clear that the role of vitamin C is to stabilize collagen, and contribute to tumor cell toxicity, through one or more of the mechanisms illustrated in the previous sections.

Experiments on Gulo-knockout (GULO-KO) mice challenged with murine B16FO cancer cells show that vitamin C–supplemented mice developed smaller tumors with more collagen encapsulation and fibrous capsule inter digitation. On the contrary, Gulo-KO mice deprived of ascorbate hosted large tumors with poorly defined borders showed more necrosis and mitosis, [101], thus reinforcing the notion that vitamin C plays a central role in the prevention and control of tumor growth, progression, and metastatic spread.

To further confirm their data, the author showed that vitamin C-supplemented GULO-KO mice injected with B16FO melanoma cells demonstrated a significant reduction (by 71%, $p = 0.005$) in tumor metastasis compared to GULO-KO mice on the control diet [102].

2.6. The antitoxic/chemopreventive pathway

"*Ascorbic acid is a potent detoxicant which counteracts and neutralizes the harmful effects of many poisons in the body. It will combat various inorganic poisons, such as mercury and arsenic, and it neutralizes the bad reactions of many organic poisons, drugs, and bacterial and animal toxins. Ascorbic acid detoxifies carbon monoxide, sulphur dioxide, and carcinogens, so it is the only immediate protection we have against the bad effects of air pollution and smoking*" [3].

This sentence defines yet another way through which vitamin C can prevent and fight off cancer, i.e., by neutralizing chemical carcinogens. Whether this effect depends on its antioxidant, anti-inflammatory effect, or yet other mechanisms, it is clear that vitamin C is an essential factor in cancer chemoprevention, and this cancer-preventive capacity is more likely associated with its protective effect against oxidative stress mediated by ROS [103].

Free radicals and other ROS are molecular species capable of independent existence that contains an unpaired electron in an atomic orbital. Free radicals derive either from normal metabolic processes or from external sources, such as exposure to X-rays, ozone, cigarette smoking, air pollutants, and industrial chemicals.

An antioxidant is a molecule stable enough to donate an electron to a rampaging free radical and neutralize it, thus reducing its capacity to damage. Antioxidants delay or inhibit cellular damage mainly through their free radical scavenging property.

Given this peculiarity, vitamin C, as noted by Stone, can combat the toxic effects of different organic and inorganic poisons, thus preventing and fighting cancer due to environmental pollutants.

Recently, the US Department of Agriculture and the National Cancer Institute recommend the consumption of a minimum of five servings of fruit and vegetables to prevent cancer, with vitamin C being able to reduce the risk of stomach, mouth, pharynx, esophagus, lung, pancreas, and cervical cancers [104]. Furthermore, both epidemiologic and observational studies based on food intake provide evidence for a strong, protective role of vitamin C against cancer [105, 106].

Different studies show that the vitamin C is capable of preventing nitrosation and interfere with experimentally induced carcinogenesis [107, 108].

Vitamin C administered together with diethylnitrosamine (NDEA) shows an inhibitory effect on the experimental esophageal carcinogenesis in Wistar rats [109]. It also protects against the toxic effects of a number of pesticides/insecticides, including malathion [110], imidacloprid [111], endosulfan [112], dimethoate [113], fenvalerate [114], and many others.

More importantly, vitamin C reduces the toxic effects of different anticancer agents, including cisplatin [115, 116], cyclophasphamide [117], selenium-cisplatin conjugates [118] radiation [119], arsenic [120], doxorubicin [121], alkylating agents [122], and many others.

2.7. The "adjuvant" anticancer effect

As shown in the previous section, vitamin C protects normal cells from the oxidative, genotoxic effects of chemotherapeutic agents, but this does not imply that it counteracts the cytotoxic effects of cancer chemotherapy and radiotherapy.

Regarding the clinical use of high doses of vitamin C in combination with standard anticancer chemotherapy, for a long time, and even today, detractors of vitamin C (that are still a multitude, within the scientific community!) warn against the risk that antioxidants (such as vitamin C) may enhance cancer cell proliferation.

The role of micronutrients with antioxidant properties (including vitamin C) as a useful adjunct to conventional chemotherapy and /or radiotherapy has been controversial, essentially because they could protect cancer cells from the deleterious effects of free radicals generated by the therapy, thereby preventing cancer cell death.

After an exhaustive review of the literature, encompassing more than 44 scientific articles on the effectiveness of vitamin C alone, or with other vitamins, with chemotherapy, the authors concluded that: "*... antioxidants [including vitamin C] do not protect cancer cells against free radical and growth-inhibitory effects of standard therapy. On the contrary, they enhance its growth-inhibitory effects on tumour cells, but protect normal cells against its adverse effects*" [123]. This literature review suggests that the use of vitamin C alone with chemotherapy results in increased survival, enhancement of chemotherapy, inhibition of tumor growth, decrease in the overall toxicity, modulation of genotoxicity linked to chemotherapy, distinct potentiating effect of chemotherapy, improved quality of life, and a whole series of other positive effects on the outcome of treated patients.

Recent evidence suggests that vitamin C can efficiently

- aid low-dose methotrexate (MTX) in inducing cell death in Hep3B cells [124],
- synergize arsenic trioxide in acute promyelocytic leukemia [26],
- improve chemosensitivity of ovarian cancer, reducing, at the same time, the toxicity of chemotherapy [125],
- sensitize tumor cells toward cytostatic drugs [126], and
- improve the quality of life of patients undergoing chemo/radiotherapy [127, 128].

In a position paper published in the *Journal of the American College of Nutrition* in 2001, the authors highlight that "... none of the published data on the effect of antioxidants in combination with radiation or chemotherapeutic agents on tumour cells supports this hypothesis" [129]. As the authors observe, normal and tumor cells differ in their responses to antioxidants; low-dose and high-dose antioxidants differ in their effect on tumor cells, some actions of antioxidants on tumor cells are unrelated to scavenging of free radicals, and antioxidants have profound effects on the regulation of gene expression in tumor cells.

3. What to do next

As we have seen, the expectations generated from *in vitro* and animal studies on vitamin C still wait for the confirmation of clinical studies. The discrepancy between *in vitro* and *in vivo* results is due to several factors presented in the next.

3.1. The confusion about the dose

Among the many mystifying (and sometimes pseudoscientific) data regarding the anticancer effects of vitamin C, the most blatant is surely the one concerning the dose used. While it is clear that therm "high" designates doses of vitamin C (generally administered by intravenous injection) leading to plasma concentrations in the order of millimoles (from one to several), current "institutional" clinical trials pass off as "high" doses of vitamin C of 1000 mg (1 g)! This is the case of a number of clinical trials, among which we could mention:

- A phase II trial of arsenic trioxide and ascorbic acid with temozolomide in patients with metastatic melanoma with or without central nervous system metastases [130],
- A clinical experience on the combination of arsenic trioxide and ascorbic acid in patients with refractory metastatic colorectal carcinoma [131],
- A phase I study on combination of decitabine, arsenic trioxide, and ascorbic acid for the treatment of myelodysplastic syndrome and acute myeloid leukemia [132], and many others. Clinical trials designed as the ones reported above are of no value in verifying the role of high doses of vitamin C in the treatment of cancer, since 1000 mg of the nutrient, even if administered in vein, is not a high ("pharmacologic") dose.

3.2. The level of tissue oxygenation

If we assume that one of the main mechanisms through which vitamin C in pharmacologic doses is toxic to cancer cells is the production of H_2O_2, then oxygen becomes a fundamental part of the cytotoxic activity of vitamin C against cancer. Solid tumors often contain areas subjected to acute or chronic hypoxia. Although severe or prolonged hypoxia is deleterious, adaptation to a hypoxic microenvironment, allows cancer cells to survive and proliferate in a hostile milieu. More importantly, since cell culture experiments are usually performed in an oxygen-rich environment, while solid tumors usually show a very low content of oxygen, this difference in oxygen content may explain the different outcome in vitamin C cancer cell killing *in vitro*, compared to what happens *in vivo* [133]. Overcoming cancer hypoxia may therefore represent one of the main ways to improve the anticancer activity of high doses of vitamin C in clinical settings, as commonly realized with either hyperbaric oxygen (HBO) or ozonated autohemotherapy.

3.3. The pharmaceutical preparation

Sodium ascorbate, rather than ascorbic acid, may be the preferred preparation for intravenous injection. Ascorbic acid produces a very acidic solution, when dissolved in water or saline solution, and, as such, unsuitable for intravenous injection. Therefore, in order to obtain a neutral solution, it is necessary to buffer it with either sodium bicarbonate or sodium hydroxide. However, adding a buffer may represent a major problem, in terms of stability of the solution, and therefore the sodium salt of vitamin C, which produces a pH of around 7.0, is clearly preferable.

3.4. The administration schedule

According to some author, a constant "flow" of vitamin C in the blood works as if the body would produce the nutrient on its own (the "dynamic flow" hypothesis) [134]. The slow, constant infusion of vitamin C is the best option to maintain a stable plasma level of the nutrient, by intravenous injection [135] even though this approach is not common in the treatment of cancer patients with intravenous high doses of the nutrient. In fact, the great majority of the cancer clinical trials performed so far with intravenous vitamin C use the infusion of vitamin C on alternate days, withdrawing the treatment during the weekend. With this treatment modality, systemic conditioning (the accelerated metabolism or disposal of ascorbic acid) may occur after prolonged supplementation of high doses of vitamin C. Thus, if vitamin C supplementation were to cease abruptly, the accelerated disposal of the nutrient may create a deficiency state ("rebound scurvy"), and this may represent a serious inconvenience, when treating cancer patients. This is why, since Klenner's experience with multiple administration routes, it may turn out to be useful to combine both intravenous and oral administration of large doses of vitamin C [7].

3.5. The glucose-ascorbate antagonism (GAA)

John Ely first proposed the glucose-ascorbate antagonism (GAA) theory in the 1970s [136]. According to this theory, the chemical structure of vitamin C and glucose is very similar and therefore they compete for the same transport system to enter the cells. As a consequence, both vitamin C and its oxidized form, DHAA, transported into different cell types (including adipocytes, erythrocytes, granulosa cells, neutrophils, osteoblasts, and smooth muscle cells), are inhibited by high blood glucose. Although *in vivo* studies are missing, investigations on diabetic patients have confirmed the theory. Therefore, given the inverse relationship between glucose and vitamin C blood levels, maintaining blood glucose levels within the normal range may greatly enhance the anticancer effect of vitamin C.

4. The efficacy of vitamin C in high doses against cancer: the facts

As a necessary premise to an evaluation of the anticancer properties of vitamin C, a realistic look at the state of the art on cancer chemotherapy can be helpful. The "war on cancer," officially declared by President Richard Nixon, with the National Cancer Act, in 1971, has been largely considered a failure, by the experts [137, 138], because of the following:

- The major improvements in survival rates mainly concern cancers of children and young adults, which account for 1.3% of all known cancers, and this has a little impact on the overall picture. Therefore, in most cases and for most forms of cancer, the war (the "war" metaphor) has been lost [139, 140].
- Targeted therapies (the "magic bullet" metaphor) are not curative because cancer usually adapts itself, becoming resistant to every new "weapon" used [141].

- Since the overwhelming majority of cancer is due to environmental, particularly lifestyle, factors, prevention, rather than cure, should be the foremost aim [142].
- The industry continues to be developed and the institutional organisms approve new cancer drugs, based on marginal improvements in survival at an unsustainably high financial cost [143].
- Furthermore, cancer chemotherapy has an inherent toxicity, which, in many instances, encompasses, among others, nausea, vomiting, mucositis, hair loss, bone marrow toxicity, cardiac, neurologic, and renal toxicity, and, in the long term, sterility and secondary malignancy.
- Finally, both early and recent reports demonstrate that cancer chemotherapy can be either ineffective/useless [144, 145] or definitely harmful [146, 147].

Compared to the current chemotherapeutic agents, vitamin C in high ("pharmacologic") concentration has the following advantages:

- It is a natural compound that is usually produced by the vast majority of plants and animals, but not (no longer!) by humans. As any other natural product, it is neither patentable nor commercially exploitable.
- Both tissues and plasma of cancer/leukemia patients show reduced levels of this nutrient and therefore the routine administration of adequate amounts of vitamin C to these patients is not only warranted, but highly desirable.
- High concentrations of vitamin C within tumor cells are associated with extended disease-free survival, while low concentrations are associated with aggressive tumor phenotype.
- It has no relevant side effects, with the exception of a slight diarrhea, a "guiding symptom" that indicates that the body "saturation" with the nutrient ("bowel tolerance"). To control the mild diarrhea that follows the body saturation threshold, it is sufficient to reduce or fractionate the doses, but in clinical practice, it is useful to maximize the effects of vitamin C, when assumed by mouth. Given the almost total absence of side effects, undue and sometimes laughable attempts to warn against its use have come, from time to time, from detractors of vitamin C, dealing with the possibility that several grams per day could lead to oxalate stone formation. In this regard, it will suffice to mention the European Food Safety Authority (EFSA) report n. EFSA-Q-2003-018, which, on this matter, clearly affirms: *"No significant relationships were found in an analysis of data from 5214 men and 5785 women between serum vitamin C concentrations and the prevalence of kidney stones"* [148]. Moreover, the hyperoxaluria associated with the use of high-dose vitamin C is primarily due to a laboratory artifact, resulting from the conversion of vitamin C to oxalate *ex vivo* (i.e., after it has left the body, while it is in the collection bottle) [149]. The only contraindication to the treatment with high doses of intravenous vitamin C is the deficiency of glucose-6-phosphate dehydrogenase (G6-P-D), a rare genetic disorder in which a number of different drugs are usually contraindicated.
- It is inexpensive.

Beyond all these advantages, vitamin C in high doses is clearly cytotoxic for a large number of human tumor cell lines. At plasma concentrations achieved by intravenous administration, vitamin C induces death in 75% of 48 cancer cell lines tested *in vitro* [150], but has no toxic effect on human peripheral white blood cells, fibroblasts, or epithelial cells. This represents the realization of the dream of the "magic bullet," even though the "scientific community" seems to continue to ignore it!

Regarding the anticancer efficacy of vitamin C *in vivo*, although the results of the first clinical trials have been rather disappointing, an unbiased analysis of the data currently available reveals an excellent safety profile, a clear-cut improvement of the quality of life [127], and a potentially important antitumor activity even though further, well-designed, controlled studies are strongly required.

5. Still looking for a "rationale": is too much of it good for nothing?

Almost 50 years after the discovery of the anticancer properties of vitamin C, scientists are still looking for a rationale for the use of this nutrient in the treatment of cancer. However, as we have seen, this rationale not only exists, but it is also evidence-based, well-founded, complex, and variegated, given the many extraordinary benefits of vitamin C for human health. Therefore, faced with such an overwhelming evidence in favor of the efficacy of vitamin C against cancer, the question may become "why mega doses of vitamin C have not yet entered the routine clinical treatment of cancer?"

As sad as it may appear, the many advantages of vitamin C as an anticancer agent represent likewise limitations to its use in clinical practice. In fact, vitamin C is a natural compound, and this implies that no pharmaceutical company can effectively exploit it for commercial purposes. Drug companies must patent the molecular structure of the active ingredient of their products in order to make a profit. Natural substances, such as vitamin C, cannot undergo any patent submission procedure, because they exist in nature.

Another important issue may be the price: compared to the high costs of cancer drugs, some of which may reach the 30,000 USD for a single dose [151], vitamin C with its price ranging from 20 to 40 USD per kilogram (depending on the country) represents a real outsider within such an expensive market. No pharmaceutical company would ever invest in the clinical development of such an inexpensive product!

Going deeper into this apparent "lack of interest" for vitamin C as an anticancer compound, we can find the outstanding issue of its selectivity of action (the "magic bullet" principle); an aspect that still fascinates the clinical oncologists. As we have seen, contrary to the chemotherapeutic agents, vitamin C kills cancer cells by exploiting a substantial metabolic difference between them and their normal counterpart; a property virtually unknown to the vast majority of the chemotherapeutic agents of common use in clinical practice!

The ethical implications of the above considerations are clear, and we will not discuss them herein. However, there are aspects, in this incomprehensible "indifference" to vitamin C as an anticancer molecule, which go far beyond plausibility and common sense.

One is surely the evidence that vitamin C deficiency is common in patients with advanced cancer, and, at the same time, patients with low plasma concentrations of vitamin C almost invariably show shorter survival, if compared to those with normal/higher concentrations [152]. Should not this evidence alone compel the clinicians to use vitamin C supplementation in cancer patients on a routine basis? A vitamin C deficiency, as we have seen in the first section of this chapter, is most probably in play in the genesis and development of cancer, and traces of this deficiency often remain, unless patients use supplements on their own, in blood and tissues of affected individuals. Is not this data alone sufficient to warrant the routine administration of vitamin C to cancer patients?

The other, not less relevant, aspect is the safety of vitamin C. The LD50 for a mouse (who normally produces its own vitamin C), is more than 3.3 g/kg of body weight, but most probably, even more than that, for mammals not producing their own vitamin, such as humans. This has clearly emerged from clinical studies on intravenous injection of mega doses of the nutrient, which have also showed a definite and unequivocal improvement in the quality of life of the treated patients.

The last aspect, regarding the importance of vitamin C in the treatment of cancer, is the demonstration of the capacity of this nutrient to reduce the side effects and improve the anticancer activity of conventional chemotherapeutic agents, when combined with them [125].

Do we really need more information to introduce high doses of vitamin C in the routine treatment of cancer?

Acknowledgements

This chapter has been realized, in part, thanks to the financial support of the "Pescara Cell Factory Foundation," and the "Associazione Italiana dei Genitori dei bambini affetti da Retinoblastoma" (AIGR).

Author details

Domenico Mastrangelo[1]*, Lauretta Massai[1], Giuseppe Fioritoni[2] and Francesco Lo Coco[3]

*Address all correspondence to: mastrangelod10@gmail.com

1 Department of Medical, Surgical and Neurological Sciences, University of Siena, Polo Scientifico San Miniato, Siena, Italy

2 Pescara Cell Factory Foundation Onlus, Corso Vittorio Emanuele, Pescara, Italy

3 Department of Biomedicine and Prevention, University of Rome Tor Vergata, Rome, Italy

References

[1] Benade L, Howard T, Burk D. Synergistic killing of Ehrlich ascites carcinoma cells by ascorbate and 3-Amino-1, 2, 4-triazole. Oncology. 1969;**23**:33-43

[2] Panel on Dietary Antioxidants and Related Compounds; Subcommittee on Upper Reference Levels of Nutrients; Subcommittee on Interpretation and Uses of DRIs; Standing Committee on the Scientific Evaluation of Dietary Reference Intakes; Food and Nutrition Board; Institute of Medicine. Dietary Reference Intakes for Vitamin C, Vitamin E, Selenium, and Carotenoids. 2000; 529 pages, ISBNs: Paperback: 978-0-309-06935-9; Hardcover: 978-0-309-06949-6. DOI:https://doi.org/10.17226/9810

[3] Stone I. The Healing Factor. "Vitamin C" Against Disease. New York: Grosset and Dunlap Inc.; 1972

[4] Stone I. Eight decades of scurvy. The case history of a misleading dietary hypothesis. Orthomolecular Psychiatry. 1979;**8**:58-62

[5] Chatterjee IB. Evolution and the biosynthesis of ascorbic acid. Science. 1971;**182**:1271-1272

[6] Stone I. The genetics of scurvy and the cancer problem. Journal of Orthomolecular Psychiatry. 1976;**5**:183-190

[7] Klenner F. The treatment of poliomyelitis and other virus diseases with vitamin C. Southern Medicine and Surgery. 1949;**111**:209-214

[8] Jungeblut CW. Further observations on vitamin C therapy in experimental poliomyelitis. Journal of Experimental Medicine. 1937;**66**:459-477

[9] Pauling L. Are recommended daily allowances for vitamin C adequate? Proceedings of the National Academy of Sciences of the United States. 1974;**71**:4442-4446

[10] Cameron E, Campbell A. The orthomolecular treatment of cancer II: Clinical trial of high dose ascorbic acid supplement in advanced human cancer. Chemico-Biological Interactions. 1974;**9**:285-315

[11] Cameron E, Pauling L. Supplemental ascorbate in the supportive treatment of cancer: Prolongation of survival times in terminal human cancer. Proceedings of the National Academy of Sciences of the United States. 1976;**73**:3685-3689

[12] Cameron E, Pauling L. Supplemental ascorbate in the supportive treatment of cancer: Reevaluation of prolongation of survival times in terminal human cancer. Proceedings of the National Academy of Sciences of the United States. 1978;**75**:4538-4542

[13] Cameron E, Pauling L. Cancer and Vitamin C. New York: W. W. Norton & Co.; 1979. [Updated and Expanded Cancer and Vitamin C, Cameron E, Pauling L, Camino Books, Inc., Phila., PA 19102; 1993]

[14] Murata A, Morishige F, Yamaguchi H. Prolongation of survival times of terminal cancer patients by administration of large doses of ascorbate. International Journal for Vitamin and Nutrition Research Supplement. 1982;**23**:103-113

[15] Creagan ET, Moertel CG, O'Fallon JR, Schutt AJ, O'Connell MJ, Rubin J, et al. Failure of high-dose vitamin C (ascorbic acid) therapy to benefit patients with advanced cancer. The New England Journal of Medicine. 1979;**301**:687-690

[16] Moertel CG, Fleming TR, Creagan ET, Rubin J, O'Connell MJ, Ames MM. High-dose vitamin C versus placebo in the treatment of patients with advanced cancer who have had no prior chemotherapy. The New England Journal of Medicine. 1985;**312**:137-141

[17] Yun J, Mullarky E, Lu C, Bosch KN, Kavalier A, Rivera K, et al. Vitamin C selectively kills KRAS and BRAF mutant colorectal cancer cells by targeting GAPDH. Science. 2015;**350**:1391-1396

[18] Van der Reest J, Gottlieb E. Anti-cancer effects of vitamin C revisited. Cell Research. 2016;**26**:269-270

[19] Reczek CR, Chandel NS. Revisiting vitamin C and cancer: A high dose of vitamin C kills certain colon cancer cells. Science. 2016:**350**:1317-1318

[20] Iqbal K, Khan A, Khattak MMAK. Biological significance of ascorbic acid (vitamin C) in human health. A review. Pakistan Journal of Nutrition. 2004;**3**:5-13

[21] Chen Q, Espey MG, Sun AY, Lee JH, Krishna MC, Shacter E, et al. Ascorbate in pharmacologic concentrations selectively generates ascorbate radical and hydrogen peroxide in extracellular fluid *in vivo*.Proceedings of the National Academy of Sciences of the United States. 2007;**104**:8749-8754

[22] Chen Q, Espey MG, Krishna MC, Mitchell JB, Corpe CP, Buettner G, et al. Pharmacologic ascorbic acid concentrations selectively kill cancer cells: Action as a pro-drug to deliver hydrogen peroxide to tissues. Proceedings of the National Academy of Sciences of the United States. 2005;**102**:13604-13609

[23] Uetaki M, Tabata S, Nakasuka F, Soga T, Tomita M. Metabolomic alterations in human cells by vitamin C-induced oxidative stress. Scientific Reports. 2015;**5**:1-9

[24] Mastrangelo D, Massai L, Micheli L, Muscettola M, Cevenini G, Grasso G. High doses of ascorbate kill Y79 retinoblastoma cells in vitro. Journal of Clinical & Experimental Ophthalmology. 2013;**4**:268. DOI: 10.4172/2155-9570.1000268

[25] Mastrangelo D, Massai M, Valyi-Nagy K, Muscettola M, Aglianò M, Lodi L, et al. Megadoses of ascorbate as a new chemotherapeutic approach in uveal melanoma: A preliminary *in vitro* investigation. Journal of Clinical & Experimental Ophthalmology. 2013;**4**:6. DOI: org/10.4172/2155-9570.1000315

[26] Mastrangelo D, Massai L, Fioritoni G, Iacone A, Di Bartolomeo P, Accorsi P, et al. Megadoses of sodium ascorbate efficiently kill HL60 cells in vitro: Comparison with arsenic trioxide. Journal of Cancer Therapeutics and Research. 2013;**4**:1366-1372

[27] Mastrangelo D, Massai L, Lo Coco F, Noguera NI, Borgia L, Fioritoni G et al. Cytotoxic effects of high concentrations of sodium ascorbate on human myeloid cell lines. Annals of Hematology. 2015;**94**:1807-1816

[28] Holman RA. A method of destroying a malignant rat tumour *in vivo*. Nature. 1957;**179**:1033

[29] Doskey CM, Buranasudja V, Wagner BA, Wilkes JG, Duc J, Cullen JJ, et al. Tumor cells have decreased ability to metabolize H_2O_2: Implications for pharmacological ascorbate in cancer therapy. Redox Biology. 2016;**10**:274-284

[30] Hacisevki A. An overview of ascorbic acid biochemistry. Journal of the Faculty of Pharmacy of Ankara. 2009;**38**:233-255

[31] Carr AC, Frei B. Does vitamin C act as pro-oxidant under physiological conditions? FASEB Journal. 1999;**13**:1007-1024

[32] Lennicke C, Rahn J, Lichtenfels R, Wessjohann LA, Seliger B. Hydrogen peroxide-production, fate and role in redox signaling of tumor cells. Journal of Cell Communication and Signaling. 2015;**13**:39. DOI: 10.1186/s12964-015-0118-6

[33] López-Lázaro M. Dual role of hydrogen peroxide in cancer: Possible relevance to cancer chemoprevention and therapy. Cancer Letters. 2007;**252**:1-8

[34] Mojić M, Bogdanović Pristov J, Maksimović-Ivanić D, Jones DR, Stanić M, Mijatović S, et al. Extracellular iron diminishes anticancer effects of vitamin C: An *in vitro* study. Scientific Reports. 2014;**4**:5955. DOI: 10.1038/srep05955

[35] Du J, Cullen JJ, Buettner GR. Ascorbic acid: Chemistry, biology and the treatment of cancer. Biochimica et Biophysica Acta. 2012;**1826**:443-457

[36] Dettman I. Iron and Vitamin C. Biological Therapies Newsletter. 2003; 1-8

[37] Hininger I, Waters R, Osmana M, Garrel C, Fernholz K, Roussel AM, et al. Acute prooxidant effects of vitamin C in EDTA chelation therapy and long-term antioxidant benefits of therapy. Free Radical Biology & Medicine. 2005;**38**:1565-1570

[38] Klenner F.Observations on the dose and administration of ascorbic acid when employed beyond the range of a vitamin in human pathology. Journal of Applied Nutrition.1971;**23**:3&4

[39] Loenarz C, Schofield CJ. Physiological and biochemical aspects of hydroxylations and demethylations catalyzed by human 2-oxoglutarate oxygenases. Trends in Biochemical Sciences. 2011;**36**:7-18. DOI: 10.1016/j.tibs.2010.07.002

[40] Ozer A, Bruick RK. Non-heme dioxygenases: Cellular sensors and regulators jelly rolled into one? Nature Chemical Biology. 2007;**3**:144-153. DOI: 10.1038/nchembio863

[41] Aprelikova O, Chandramouli GVR, Wood M, Vasselli JR, Riss J, Maranchie JK, et al. Regulation of HIF Prolyl hydroxylases by hypoxia-inducible factors. Journal of Cellular Biochemistry. 2004;**92**:491-501

[42] Schofield CJ, Ratcliffe PJ. Oxygen sensing by HIF hydroxylases. Nature Reviews Molecular Cell Biology. 2004;**5**:343-354

[43] Knowles HJ, Raval RR, Harris AL, Ratcliffe PJ. Effect of ascorbate on the activity of hypoxia-inducible factor in cancer cells. Cancer Research. 2003;**63**:1764-1768

[44] Tian W, Wang Y, Xu Y, Guo X, Wang B, Sun L, et al. The hypoxia-inducible factor renders cancer cells more sensitive to vitamin C-induced toxicity. Journal of Biological Chemistry. 2014;**289**:3339-3351

[45] Kuiper C, Vissers MCM. Ascorbate as a co-factor for Fe- and 2-oxoglutarate dependent dioxygenases: Physiological activity in tumor growth and progression. Frontiers in Oncology. 2014;**4**:1-11

[46] Kuiper C, Molenaar IG, Dachs GU, Currie MJ, Sykes PH, Vissers MC. Low ascorbate levels are associated with increased hypoxia-inducible factor-1 activity and an aggressive tumor phenotype in endometrial cancer. Cancer Research. 2010;**70**:5749-5758

[47] Kuiper C, Dachs GU, Munn D, Currie MJ, Robinson BA, Pearson JF. Increased tumor ascorbate is associated with extended disease-free survival and decreased hypoxia-inducible factor-1 activation in human colorectal cancer. Frontiers in Oncology. 2014;**4**:1-10

[48] Semenza GL. Hypoxia-inducible factors: Mediators of cancer progression and targets for cancer therapy. Trends in Pharmacological Sciences. 2012;**33**:207-214

[49] Muz B, de la Puente P, Azab F, Azab AK. The role of hypoxia in cancer progression, angiogenesis, metastasis, and resistance to therapy. Hypoxia (Auckl). 2015;**11**:83-92

[50] Masoud GN, Li W. HIF-1α pathway: Role, regulation and intervention for cancer therapy. Acta Pharmaceutica Sinica B. 2015;**5**:378-389

[51] Kawada H, Kaneko M, Sawanobori M, Uno T, Matsuzawa H, Nakamura Y, et al. High concentrations of l-ascorbic acid specifically inhibit the growth of human leukemic cells via downregulation of HIF-1α transcription. PLoS One. 2013;**8**:e62717

[52] Gao P, Zhang H, Dinavahi R, Li F, Xiang Y, Raman V, et al. HIF-dependent antitumorigenic effect of antioxidants *in vivo*. Cancer Cell. 2007;**12**:230-238

[53] Mikirova N, Scimeca RC. Gene expression response to ascorbic acid in mice implanted with sarcoma S180 cells. Journal of Translational Science. 2016;**2**:145-153

[54] Montfort A, Wuts A. Breathing-in epigenetic change with vitamin C. EMBO Reports. 2013;**14**:337-346

[55] Tahiliani M, Koh KP, Shen Y, Pastor WA, Bandukwala H, Brudno Y, et al. Conversion of 5-methylcytosine to 5-hydroxymethylcytosine in mammalian DNA by MLL partner TET1. Science. 2009;**324**:930-935. DOI: 10.1126/science.1170116. PMC 2715015Freely accessible. PMID 19372391

[56] Ito S, Shen L, Dai Q, Wu SC, Collins LB, Swenberg JA, et al. Tet proteins can convert 5-methylcytosine to 5-formylcytosine and 5-carboxylcytosine. Science. 2011;**333**:1300-1303. DOI: 10.1126/science.1210597. PMC 3495246Freely accessible. PMID 21778364

[57] Camarena V, Wang G. The epigenetic role of vitamin C in health and disease. Cellular and Molecular Life Sciences. 2016;**73**:1645-1658. DOI: 10.1007/s00018-016-2145-x

[58] Young JI, Züchner S, Wang G. Regulation of the epigenome by vitamin C. Annual Review of Nutrition. 2015;**35:**545-564. DOI: 10.1146/annurev-nutr-071714-034228

[59] Chung TL, Brena RM, Kolle G, Grimmond SM, Berman BP, Laird PW et al. Vitamin C promotes widespread yet specific DNA demethylation of the epigenome in human embryonic stem cells. Stem Cells. 2010;**28**:1848-1855

[60] He XB, Kim M, Kim SY, Yi SH, Rhee YH, Kim T, et al. Vitamin C facilitates dopamine neuron differentiation in fetal midbrain through TET1 and JMJD3-dependent epigenetic control manner. Stem Cells. 2015;**33**:1320-1332

[61] Gao Y, Han Z, Li Q, Wu Y, Shi X, Ai Z, et al. Vitamin C induces a pluripotent state in mouse embryonic stem cells by modulating microRNA expression. FEBS Journal. 2015;**282**:685-699

[62] Blaschke K, Ebata KT, Karimi MM, Zepeda-Martınez JA, Goyal P, Mahapatra S, et al. Vitamin C induces Tet-dependent DNA demethylation and a blastocyst-like state in ES cells. Nature. 2013;**500**:222-228

[63] Sajadian SO, Tripura C, Samani FS, Ruoss M, Dooley S, Baharvand H, et al. Vitamin C enhances epigenetic modifications induced by 5-azacytidine and cell cycle arrest in the hepatocellular carcinoma cell lines HLE and Huh7. Clinical Epigenetics. 2016;**8**:46

[64] Liu M, Ohtani H, Zhou W, Due Ørskov A, Charlet J, Zhang YW, et al. Vitamin C increases viral mimicry induced by 5-aza-2′-deoxycytidine. Proceedings of the National Academy of Sciences of the United States. 2016;**113**:10238-10244

[65] Gustafson CB, Yang C, Dickson KM, Shao H, Booven D, Harbour JW, et al. Epigenetic reprogramming of melanoma cells by vitamin C treatment. Clinical Epigenetics. 2015;**7**:51

[66] Venturelli S, Sinnberg TW, Berger A, Noor S, Levesque MP, Böcker A, et al. Epigenetic impacts of ascorbate on human metastatic melanoma cells. Frontiers in Oncology. 2014;**4**:227. Available from: https://doi.org/10.3389/fonc.2014.00227

[67] Zhao L, Quan Y, Wang J, Wang F, Zheng Y, Zhou A. Vitamin C inhibit the proliferation, migration and epithelial-mesenchymal-transition of lens epithelial cells by destabilizing HIF-1α. International Journal of Clinical and Experimental Medicine. 2015;**8**:15155-15163

[68] Wintergerst ES, Maggini S, Hornig DH. Immune-enhancing role of vitamin C and zinc and effect on clinical conditions. Annals of Nutrition and Metabolism. 2006;**50**:85-94

[69] Sorice A, Guerriero E, Capone F, Colonna G, Castello G, Costantini S. Ascorbic acid: Its role in immune system and chronic inflammation diseases. Mini-Reviews in Medicinal Chemistry. 2014;**14**:444-452

[70] Combs Jr GF. The Vitamins. 4th ed. Academic Press; 2012. p. 598. Paperback ISBN: 9780128102442, eBook ISBN:9780123819819,Hardcover ISBN: 9780123819802

[71] Pavlovic V, Sarac M. A short overview of vitamin C and selected cells of the immune system. Central European Journal of Medicinal Chemistry. 2010;**6**:1-10

[72] Lewin S. Vitamin C. Its Molecular Biology and Medical Potential. New York, NY: Academic Press; 1976. p. 244

[73] Goetzl EJ, Wasserman SI, Gigli I, Austen KF. Enhancement of random migration and chemotactic response of human leukocytes by ascorbic acid. Journal of Clinical Investigation. 1974;**53**:813-818

[74] Dahl H, Degre M. The effect of ascorbic acid on production of human interferon and the antiviral activity *in vitro*. Acta Agriculturae Scandinavica, Section B. 1976;**84**:280-284

[75] Cho D, Hahm E, Kang JS, Kim YI, Yang YH, Park JH, et al. Vitamin C downregulates interleukin-18 production by increasing reactive oxygen intermediate and mitogen-activated protein kinase signaling in B16F10 murine melanoma cells. Melanoma Research. 2003;**13**:549-554

[76] Ehrlich P. Ueber den jetzigen Stand der Karzinomforschung. Ned Tijdschr Geneeskd. 1909;**5**:273-290

[77] Burnet FM. Immunological surveillance in neoplasia. Transplantation Reviews. 1971;**7**:3-25

[78] Dunn GP, Old LJ, Schreiber D. The immunobiology review of cancer immunosurveillance and immunoediting. Immunity. 2004;**21**:137-148

[79] Schwager J, Schulze J. Modulation of interleukin production by ascorbic acid. Veterinary Immunology and Immunopathology. 1998;**64**:45-57

[80] Pavlovic V, Pavlovic D, Kocic G, Sokolovic D, Sarac M, Jovic Z. Ascorbic acid modulates monosodium glutamate induced cytotoxicity in rat thymus. Bratislavske Lekarske Listy. 2009;**110**:205-209

[81] Wu CC, Doriarajan T, Lin TL. Effect of ascorbic acid supplementation on the immune response of chickens vaccinated and challenged with infectious bursal disease virus. Veterinary Immunology and Immunopathology. 2000;**74**:145-152

[82] Pavlovic V, Cekic S, Bojanic V, Stojiljkovic N, Rankovic G. Ascorbic acid modulates spontaneous thymocyte apoptosis. Acta Medica International. 2005;**44**:21-23

[83] Perez-Cruz I, Carcamo JM, Golde DW. Vitamin C inhibits FAS-induced apoptosis in monocytes and U937 cells. Blood. 2003;**102**:336-343

[84] Vojdani A, Bazargan M, Vojdani E, Wright J. New evidence for antioxidant properties of Vitamin C. Cancer Detection and Prevention. 2000;**24**:508-523

[85] Campbell JD, Cole M, Bunditrutavorn B, Vella AT. Ascorbic acid is a potent inhibitor of various forms of T cell apoptosis. Cellular Immunology. 1999;**194**:1-5

[86] Tan PH, Sagoo P, Chan C, Yates JB, Campbell J, Beutelspacher SC, et al. Inhibition of NF-kappa B and oxidative pathways in human dendritic cells by antioxidative vitamins generates regulatory T cells. The Journal of Immunology. 2005;**174**:7633-7644

[87] Richards E. Vitamin C and Cancer: Medicine or Politics? McMillan Professional & Academic LTD; 1991. p. 269. ISBN-13:978-0312052423, ISBN-10:0312052421

[88] Whiteside TL. Immune suppression in cancer: Effects on immune cells, mechanisms and future therapeutic intervention. Seminars in Cancer Biology. 2006;**16**:3-15

[89] Verma R, Foster RE, Horgan K, Mounsey K, Nixon H, Smalle N, et al. Lymphocyte depletion and repopulation after chemotherapy for primary breast cancer. Breast Cancer Research. 2016;**18**:10. DOI: 10.1186/s13058-015-0669-x

[90] Grossman SA, Ellsworth S, Campian J, Wild AT, Herman JM, Laheru D, et al. Survival in patients with severe lymphopenia following treatment with radiation and chemotherapy for newly diagnosed solid tumors. Journal of the National Comprehensive Cancer Network. 2015;**13**:1225-1231

[91] McCormick WJ. Cancer: A collagen disease, secondary to a nutritional deficiency. Archives of Pediatric. 1959;**76**:166-171

[92] Grosso G, Bei R, Mistretta A, Marventano S, Calabrese G, Masuelli L, et al. Effects of vitamin C on health: A review of evidence. Frontiers in Bioscience Landmark Edition. 2013;**18**:1017-1029

[93] McCormick WJ. Is cancer a collagen disease attributable to vitamin C deficiency? Union Medicale Du Canada. 1959;**88**:700-704

[94] Ikeda K, Iyama K, Ishikawa N, Egami H, Nakao M, Sado Y, et al. Loss of expression of type IV collagen 5 and 6 chains in colorectal cancer associated with the hypermethylation of their promoter region. The American Journal of Pathology. 2006;**168**:856-865

[95] Tanjore H, Kalluri R. The role of type IV collagen and basement membranes in cancer progression and metastasis. The American Journal of Pathology. 2006;**168**:715-717

[96] Roomi MW, Monterrey JC, Kalinovsky T, Rath M, Niedzwiecki A. Inhibition of invasion and MMPs by a nutrient mixture in human cancer cell lines: A correlation study. Experimental Oncology. 2010;**32**:243-248

[97] Cathcart J, Pulkoski-Gross A, Cao J. Targeting matrix metalloproteinases in cancer: Bringing new life to old ideas. Genes and Diseases. 2015;**2**:26-34

[98] Rath M, Pauling L. Plasmin-induced proteolysis and the role of apoprotein(a), lysine and synthetic analogs. Orthomolecular Medicine. 1992;**7**:17-23

[99] Roomi MW, Monterrey JC, Kalinovsky T, Rath M, Niedzwiecki A. Comparative effects of EGCG, green tea and a nutrient mixture on the patterns of MMP-2 and MMP-9 expression in cancer cell lines. Oncology Reports. 2010;**24**:747-757

[100] Niedzwiecki A, Roomi MW, Kalinovsky T, Rath M. Anticancer efficacy of polyphenols and their combinations. Nutrients. 2016;**8**(552):1-17. DOI:10.3390/nu8090552

[101] Cha J, Roomi MW, Ivanov V, Kalinovsky T, Niedzwiecki A, Rath M. Ascorbate depletion increases growth and metastasis of melanoma cells in vitamin C deficient mice. Experimental Oncology. 2011;**33**:226-230

[102] Cha J, Roomi MW, Ivanov V, Kalinovsky T, Niedzwiecki A, Rath M. Ascorbate supplementation inhibits growth and metastasis of B16FO melanoma and 4T1 breast cancer cells in vitamin C-deficient mice. International Journal of Oncology. 2013;**42**:55-64. DOI: 10.3892/ijo.2012.1712

[103] Lee KW, Lee HJ, Surh YJ, Lee CY. Vitamin C and cancer chemoprevention: Reappraisal. American Journal of Clinical Nutrition. 2003;**78**:1074-1078

[104] World Cancer Resarch Fund/American Institute for Cancer Research. Food Research and the Prevention of Cancer: A Global Perspective. Washington, DC: American Institute for Cancer Research; 1997. p. 517

[105] Vecchia CL, Altieri A, Tavani A. Vegetables, fruit, antioxidants and cancer: A review of Italian studies. European Journal of Nutrition. 2001;**40**:261-267

[106] Henson DE, Block G, Levine M. Ascorbic acid: Biologic functions and relation to cancer. Journal of the National Cancer Institute. 1991;**83**:547-552

[107] Bollschweiler E, Wolfgarten E, Nowroth T, Rosendahl U, Monig SP, Holscher AH. Vitamin intake and risk of subtypes of esophageal cancer in Germany. Journal of Cancer Research and Clinical Oncology. 2002;**128**:575-580

[108] Mirvish SS, Wallcave L, Eagen M, Shubik P. Ascorbate-nitrite reaction: Possible means of blocking the formation of carcinogenic N-nitroso compounds. Science. 1972;**177**:65-68

[109] Ramos AC, Araujo MR, Lopes LR, Andreollo NA. Role of the vitamin C in diethylnitrosamine-induced esophageal cancer in Wistar rats. Acta Cirurgica Brasileira. 2009;**24**:183-188

[110] Ismail SM. Protective effects of vitamin C against biochemical toxicity induced by malathion pesticides in male albino rat. Journal of Evolutionary Biology. 2013;**5**:1-5

[111] Eissa OS. Protective effect of vitamin C and glutathione against the histopathological changes induced by imidacloprid in the liver and testis of Japanese Quail. Egyptian Journal of Hospital Medicine. 2004;**16**:39-54

[112] Takhshid MA, Tavasuli AR, Heidary Y, Keshavarz M, Kargar H. Protective effect of vitamins E and C on endosulfan-induced reproductive toxicity in male rats. Iranian Journal of Medical Sciences. 2012;**37**:173-180

[113] Dubey S, Shrivastava R, Binjhade A, Chouhan R, Shrivastava VK. Antagonistic role of vitamin-C against dimethoate toxicity on kidney in air breathing fish Clarias batrachus (Linn.). International Journal of Current Research. 2015;**3**:25-34

[114] Hussein HK. Elnaggar MH, Al-Dailamy JM. Protective role of vitamin C against hepatorenal toxicity of fenvalerate in male rats. Garjest. 2012;**1**:060-065

[115] Appenroth D, Fröb S, Kersten L, Splinter FK, Winnefeld K. Protective effects of vitamin E and C on cisplatin nephrotoxicity in developing rats. Archives of Toxicology. 1997;**71**:677-683

[116] Antunes LM, Araujo MCP, Darin JDAC, Bianchi MLP. Effects of the antioxidants curcumin and vitamin C on cisplatin-induced clastogenesis in Wistar rat bone marrow cells. Mutation Research. 2000;**465**:131-137

[117] Ghosh S, Ghosh D, Chattopadhyay S, Debnath J. Effect of ascorbic acid supplementation on liver and kidney toxicity in cyclophosphamide-treated female albino rats. The Journal of Toxicological Sciences. 1999;**24**:141-144

[118] Blasiak J, Kowalik J. Protective action of vitamin C against DNA damage induced by selenium-cisplatin conjugate. Acta Biochimica Polonica. 2001;**48**:233-240

[119] Sato T, Kinoshita M, Yamamoto T, Ito M, Nishida T, Takeuchi M, et al. Treatment of irradiated mice with High-Dose ascorbic acid reduced lethality. PloS One. 2015;10:e0117020. DOI.org/10.1371/journal.pone.0117020. PubMed PMID: 25651298. PubMed PMCID: 4317183

[120] Mondal R, Biswas S, Chatterjee A, Mishra R, Mukhopadhyay A, Bhadra RK, et al. Protection against arsenic-induced hematological and hepatic anomalies by supplementation of vitamin C and vitamin E in adult male rats. Journal of Basic and Clinical Physiology and Pharmacology. 2016;**27**:643-652. DOI: 10.1515/jbcpp-2016-0020

[121] Ludke A, Sharma AK, Bagchi AK, Singal PK. Subcellular basis of vitamin C protection against doxorubicin-induced changes in rat cardiomyocytes. Molecular and Cellular Biochemistry. 2012;**360**:215-224

[122] Aidoo A, Lyn-Cook LE, Lensing S, Wamer W. Ascorbic acid (vitamin C) modulates the mutagenic effects produced by an alkylating agent in vivo. Environmental and Molecular Mutagenesis. 1994;**24**:220-228

[123] Stoute JO. The use of vitamin C with chemotherapy in cancer treatment: An annotated bibliography. Journal of Orthomolecular Medicine. 2004;**19**:198-245

[124] Yiang GT, Chou PL, Hung YT, Chen JN, Chang WJ, Yu YL, et al. Vitamin C enhances anticancer activity in methotrexate-treated Hep3B hepatocellular carcinoma cells. Oncology Reports. 2014;**32**:1057-1063. DOI: 10.3892/or.2014.3289. Epub Jun 25, 2014

[125] Ma Y, Chapman J, Levine M, Polireddy K, Drisko J, Chen Q. High-Dose parenteral ascorbate enhanced chemosensitivity of ovarian cancer and reduced toxicity of chemotherapy. Science Translational Medicine. 2014;**6**:222ra18

[126] Frömberg A, Gutsch D, Schulze D, Vollbracht C, Weiss G, Czubayko F, et al. Ascorbate exerts anti-proliferative effects through cell cycle inhibition and sensitizes tumor cells towards cytostatic drugs. Cancer Chemotherapy and Pharmacology. 2011;**67**:1157-1166

[127] Carr AC, Vissers MCM, Cook JS. The effect of intravenous vitamin C on cancer- and chemotherapy-related fatigue and quality of life. Frontiers in Oncology. 2014;**4**:283

[128] Stephenson CM, Levin RD, Spector T, Lis CG. Phase I clinical trial to evaluate the safety, tolerability, and pharmacokinetics of high-dose intravenous ascorbic acid in patients with advanced cancer. Cancer Chemotherapy and Pharmacology. 2013;**72**:139-146

[129] Prasad KN, Cole WC, Kumar B, Prasad KC. Scientific rationale for using High-Dose multiple micronutrients as an adjunct to standard and experimental cancer therapies. Journal of the American College of Nutrition. 2001;**20**:450S-463S

[130] Bael TE, Peterson BL, Gollob JA. Phase II trial of arsenic trioxide and ascorbic acid with temozolomide in patients with metastatic melanoma with or without central nervous system metastases. Melanoma Research. 2008;**18**:147-151. DOI: 10.1097/CMR.0b013e3282f2a7ae

[131] Subbarayan PR, Lima M, Ardalan B. Arsenic trioxide/ascorbic acid therapy in patients with refractory metastatic colorectal carcinoma: A clinical experience. Acta Oncologica. 2007;**46**:557-561

[132] Welch JS, Klco JM, Gao F, Procknow E, Uy GL, Stockerl-Goldstein KE, et al. Combination decitabine, arsenic trioxide, and ascorbic acid for the treatment of myelodysplastic syndrome and acute myeloid leukemia: A phase I study. American Journal of Hematology. 2011;**86**:796-800. DOI: 10.1002/ajh.22092. Epub Aug 3 2011

[133] Carreau A, El Hafny-Rahbi B, Matejuk A, Grillon C, Kieda C. Why is the partial oxygen pressure of human tissues a crucial parameter? Small molecules and hypoxia. Journal of Cellular and Molecular Medicine. 2011;**15**:1239-1253

[134] Hickey DS, Roberts HJ, Cathcart RF. Dynamic flow: A new model for ascorbate. Journal of the Minerals. 2005;**20**:237-244

[135] Duconge J, Miranda-Massari JR, Gonzalez MJ, Jackson JA, Warnock W, Riordan NH. Pharmacokinetics of vitamin C: Insights into the oral and intravenous administration of ascorbate. Puerto Rico Health Sciences. 2008;**27**:7-19

[136] Hamel EE, Santisteban GA, Ely JT, Read DH. Hyperglycemia and reproductive defects in non-diabetic gravidas: A mouse model test of a new theory. Life Sciences. 1986;**39**:1425-1428

[137] Haines I. The war on cancer: Time for a new terminology. Lancet. 2014;**383**:1883

[138] Hanin L. Why victory in the war on cancer remains elusive: Biomedical hypotheses and mathematical models. Cancers. 2011;**3**:340-367

[139] Sporn BM. The war on cancer. Lancet. 1996;**347**:1377-1381

[140] Temple NJ, Burkitt DP. The war on cancer-failure of therapy and research: Discussion paper. Journal of the Royal Society of Medicine. 1991;**84**:95-98

[141] Huang S. The war on cancer: Lessons from the war on terror. Frontiers in Oncology. 2014;**4**:1-4

[142] Hanahan D. Rethinking the war on cancer. Lancet. 2014;**383**:558-563

[143] Mailankody S, Prasad V. Comparative effectiveness questions in oncology. The New England Journal of Medicine. 2014;**370**:1478-1481. DOI: 10.1056/NEJMp1400104

[144] Abel U. Chemotherapy of advanced epithelial cancer-a critical review. Biomedicine & Pharmacotherapy. 1992;**46**:439-452

[145] Morgan G, Ward R, Barton M. The contribution of cytotoxic chemotherapy to 5-year survival in adult malignancies. Clinical Oncology Journal (The Royal College of Radiologists). 2004;**16**:549-560

[146] Wise PH. Cancer drugs, survival, and ethics. British Medical Journal. 2016;**355**:i5792

[147] Godlee F. Too much chemotherapy. British Medical Journal. 2016;**355**:i6027

[148] Opinion of the Scientific Panel on Dietetic Products, Nutrition and Allergies on a request from the Commission related to the Tolerable Upper Intake Level of Vitamin C (L-Ascorbic acid, its calcium, potassium and sodium salts and L-ascorbyl-6-palmitate). The EFSA Journal. 2004;**59**:1-21

[149] Gaby AR. Nutritional Medicine. Fritz Perlberg Publishing; 2011. p. 1358. ISBN-13:978-0982885000, ISBN-10:0982885008

[150] Chen Q, Espey MG, Sun AY, Pooput C, Kirk KL, Krishna MC, et al. Pharmacologic doses of ascorbate act as a prooxidant and decrease growth of aggressive tumor xenografts in mice. Proceedings of the National Academy of Sciences of the United States. 2008;**105**:11105-11109

[151] Siddiqui M, Rajkumar SV. The high cost of cancer drugs and what we can do about it. Mayo Clinic Proceedings. 2012;**87**(10):935-943

[152] Mayland CR, Bennett MA, Allan K. Vitamin C deficiency in cancer patients. Palliative Medicine. 2005;**19**:17-20

Chapter 5

The Role of Vitamin C in the Protection and Modulation of Genotoxic Damage Induced by Metals Associated with Oxidative Stress

María del Carmen García-Rodríguez,
Alejandro Gordillo-García and
Mario Altamirano-Lozano

Additional information is available at the end of the chapter

http://dx.doi.org/10.5772/intechopen.68686

Abstract

This chapter reviews the effects of vitamin C on metal-induced genotoxicity. By focusing on cutting-edge studies, including our own results in experiments with vanadium(V) and chromium(VI), the suggestion that vitamin C can be used effectively to protect against or reduce the genotoxic effects induced by metal exposure by suppressing oxidative stress is particularly explored. After explaining the chemical mechanisms involved in oxidative stress associated with heavy metals, this chapter discusses the various proposals regarding the physiological processes of vitamin C at the molecular level, its relationship with oxidative stress, levels of 8-hydroxydeoxyguanosine (8-OH-dG, 7,8-dihy-dro-8-oxodeoxyguanosine) and apoptosis, and its role in the protection and modulation of DNA damage, as well as how they fit with our own results that showed an increase in apoptosis and 8-OH-dG when vitamin C was administered in addition to the metallic compounds. The relevant gaps in our understanding of the role of vitamin C with regard to these issues are highlighted, as well as the key importance of its clinical use, and ultimately, human health.

Keywords: vitamin C, antigenotoxic, genotoxic damage, antioxidant, heavy metals, oxidative stress

1. Introduction

Several studies have suggested that diets rich in fresh fruits and vegetables are associated with a lower risk of cardiovascular diseases and cancer because of the high levels of antioxidants such as vitamin C and polyphenols present in these foods [1]. The antioxidant effects of vitamin C have been observed both *in vitro* and *in vivo*. Ascorbic acid, which is a water-soluble bioactive form of vitamin C (**Figure 1**), can be found in all body fluids. At physiological pH, the 99% of vitamin C is present as $AscH^-$ (**Figure 1b**), indicating that its chemical form confers its main antioxidant effects. The remaining percentage is covered by 0.05% of AscH2 (**Figure 1a**) and 0.004% of $Asc2^-$ (**Figure 1c**). The antioxidant activity of vitamin C develops in two ways: (a) directly, by scavenging oxygen free-radicals, more generally known as reactive oxygen species (ROS) and (b) indirectly, by regenerating other antioxidant systems [2, 3].

A significant number of studies have focused on metal-induced toxicity and carcinogenicity by emphasizing their role in the generation of ROS. Metal-mediated formation of free radicals may cause modifications to DNA bases, lipid peroxidation, and changes in calcium and sulfhydryl homeostasis [4, 5]. However, these effects can be influenced by the action of low molecular weight antioxidants such as vitamin C, which is capable of chelating metal ions, reducing their catalytic activity, and resulting ROS formation. Since the genotoxicity of heavy metals associated with oxidative stress is based on the oxidative mechanism during reduction [5], vitamin C can be used effectively to protect or reduce the induced genotoxic effects by suppressing oxidative stress caused by these metallic compounds [6–8]. However, paradoxically under certain conditions (i.e., low concentration *in vitro* and the presence of metal ions), vitamin C can exert a pro-oxidant effect, increasing oxidative damage to lipids, DNA and protein, besides being a potential direct or indirect modulator of gene expression [9]. In fact, our understanding of the physiological processes of vitamin C at the molecular level and its relationship with oxidative stress, as well as its role in the protection and modulation of DNA damage is still incomplete. As a consequence, the evidence indicating the potential of vitamin C in counteracting oxidative stress, a key component in various pathological conditions including cardiovascular disease, neurological disorders, diabetes, and cancer [3, 10, 11], has not been translated, at least conclusively, in many randomized controlled trials.

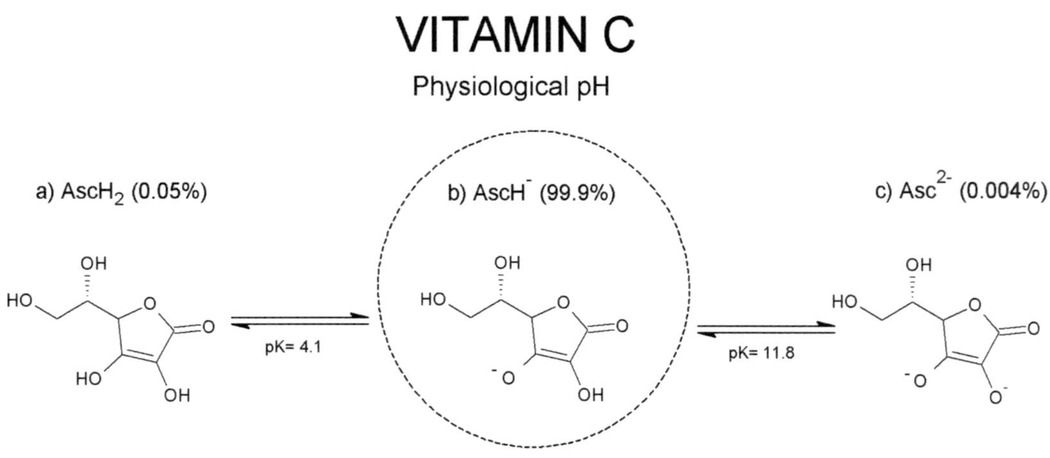

Figure 1. Bioactive forms of vitamin C at physiological pH. a) Ascorbic acid with two ionisable hydroxyl groups, AscH2; b) Ascorbate anion, $AscH^-$; and c) Ascorbate dianion, $Asc2^-$.

http://dx.doi.org/10.5772/intechopen.68686

2. Heavy metals and oxidative stress: the case of vanadium and chromium

It is well established that redox-active metals participate closely in the generation of different free radicals [6]. Exposure to transition metal ions$^{(n+)}$ such as chromium (Cr) and vanadium (V) hence represent a realistic *in vivo* production of ROS and free radicals due to intra-cellular reduction. The majority of the hydroxyl radicals (•OH) generated *in vivo* come from the metal-catalyzed breakdown of hydrogen peroxide (H_2O_2) through the Fenton and Haber-Weiss reactions [4, 12]:

$$\text{Transition metal ion}^{(n+)} + H_2O_2 \rightarrow \text{Transition metal ion}^{(n+1)} + {}^{\cdot}OH + OH^-$$

The •OH is the most reactive of all the ROS (half-life <1 ns) and interacts with all components of the DNA molecule. The initial stage of mutagenesis, carcinogenesis, and aging involves the permanent modification of genetic material. In fact, it has been well documented that in various cancer tissues, free radical-mediated DNA damage has occurred. ROS-induced DNA damage involves single- or double-stranded DNA breaks, purine, pyrimidine, or deoxyribose modifications, and DNA cross-links [5, 13, 14].

As mentioned above, the main genotoxic mechanism of V(V) and Cr(VI) compounds has been linked to reduction and generation of •OH [15, 16]. Reduction of V(V) to V(IV) takes place outside the cell (**Figure 2**). In plasma, V(V) is rapidly reduced to V(IV) by nicotinamide adenine

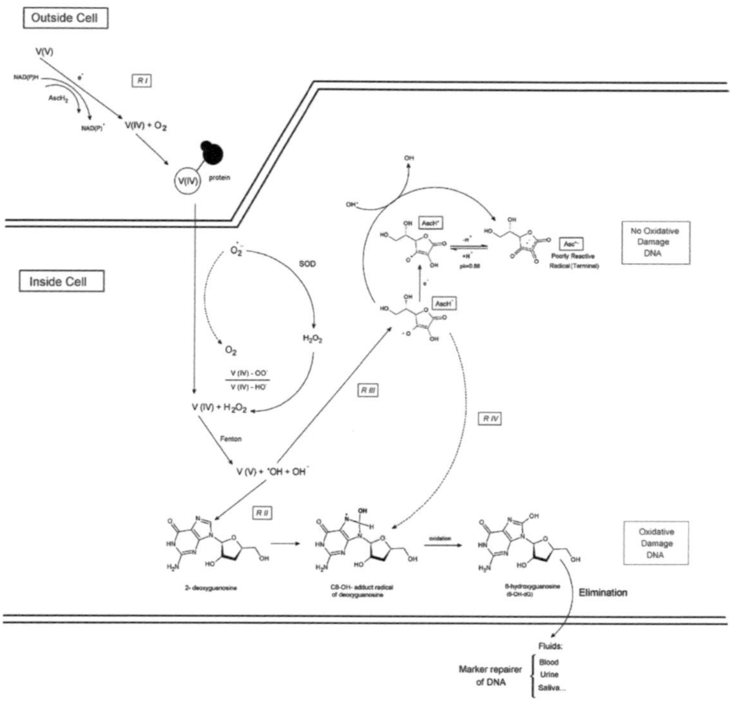

Figure 2. Routes of V(V) involved in the induction, protection and modulation of DNA oxidative damage (outside and inside cell).

dinucleotide phosphate (NADPH) and ascorbic acid. Once reduced, it is bonded with plasma proteins that carry it into the cell, where peroxovanadyl radicals [V(IV)–OO•] and vanadyl hydroperoxide [V(IV)–HO•] are formed. The generated superoxide is further converted into H_2O_2 by the dismutation reaction with superoxide dismutase (SOD). V(IV) can react through the Fenton reaction with H_2O_2 forming a •OH (**Figure 2**, *RI*) [5, 17, 18]. Nevertheless, Cr(VI) can actively enter the cells through channels for the transfer of isoelectric and isostructural anions, such as those for SO_4^{2-} and HPO_4^{2-} [19]. Once inside the cell, Cr(VI) quickly forms a complex with glutathione, reducing to Cr(V) (**Figure 3**, *RII*). Additionally, NAD(P)H can also reduce Cr(VI) to Cr(V), mediated by ascorbate (**Figure 3**, *RI*). Cr(V) can react through the Fenton reaction with H_2O_2 forming •OH [15].

The genetic damage by the production of C8-OH-adduct radical of deoxyguanosine is generated by the interaction between •OH and 2-deoxyguanosine. Therefore, there are two ways in which the protection and modulation of DNA oxidative damage could be caused. First, AscH⁻ could react with •OH, quenching and converting it into a poorly reactive semi-hydroascorbate radical, which do not cause DNA damage (**Figures 2** and **3**, *RIII*). Second, AscH⁻ can activate the repair mechanisms to eliminate C8-OH-adduct radical of deoxyguanosine. During catalysis of •OH in the reaction with 2-deoxyguanosine with molecular oxygen, C8-OH-adduct radical of deoxyguanosine is formed (**Figures 2** and **3**, *RII* and *RIV* respectively), which is a

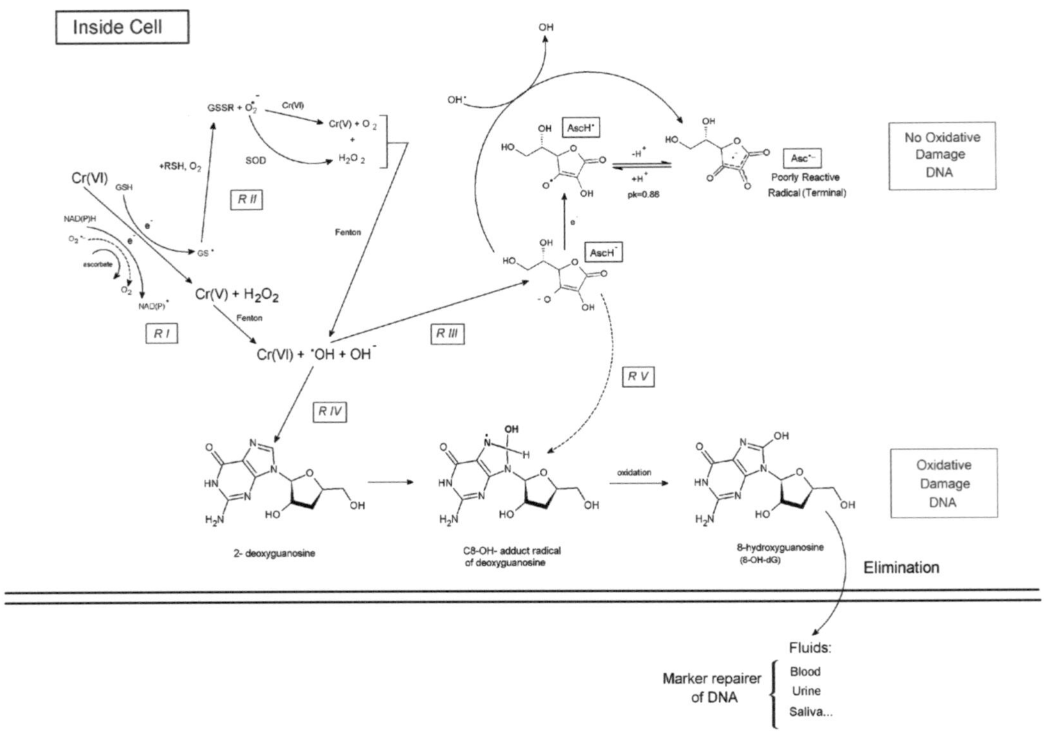

Figure 3. Routes of Cr(VI) involved in the induction, protection and modulation of DNA oxidative damage (all occurring inside the cell).

form of oxidative DNA damage because it induces DNA strand breaks [20, 21]. Thus, $AscH^-$ could activate repair mechanisms and eliminate this radical through 8-hydroxydeoxyguanosine (8-OH-dG, 7,8-dihy-dro-8-oxodeoxyguanosine), which is a marker repairer of oxidative stress in biological systems that can be measured in fluids such as blood, urine, and saliva (**Figures 2** and **3**, *RIV* and *RV*, respectively).

Although the direct relationship between DNA damage and $^{\bullet}OH$ is not completely clear, Patlolla et al. [22] have suggested a role for ROS in Cr(VI)-induced genotoxicity and cytotoxicity. They showed that Cr(VI) induced genomic DNA damage through the formation of 8-OH-dG. Nevertheless, Rudolf and Cérvinka [23] observed that Cr(VI) induced time- and concentration-dependent cytotoxicity, resulting in oxidative stress, but through suppression of antioxidant systems and by activation of p53-dependent apoptosis. Other studies have questioned the genotoxic/mutagenic effect of $^{\bullet}OH$ in the context of Cr exposure, suggesting that reduction of Cr(VI) by physiological concentrations of vitamin C generates ascorbate-Cr(III)-DNA crosslinks and binary Cr(III)-DNA adducts. Therefore, Cr-DNA adducts are responsible for both the mutagenicity and genotoxicity of Cr(VI) [24].

3. Protective effects of vitamin C against genotoxic damage from vanadium(V) and chromium(VI)

For humans, vitamin C is an essential micronutrient that plays multiple biological roles. It must be obtained from the ingestion of particular foods, mainly fresh fruits and vegetables, since our body is incapable of synthesizing it. The consequences of the intake of very high doses of vitamin C (>2 g/day) remain a subject of intense debate. However, it has been observed that supplementation of vitamin C reduces the incidence of stomach, lung and colorectal cancer; likewise, low serum levels of vitamin C in high-risk populations may contribute to increased risk of gastric metaplasia or chronic gastritis, which are both precancerous lesions [5, 25]. Nevertheless, analyses of the effects of vitamin C are rather complicated because diet and vitamin supplementation determine the levels of vitamin C in plasma.

Cameron and Pauling highlighted the beneficial properties of vitamin C in the 1970s. They suggested that high doses of vitamin C (>10 g/day) cure and prevent cancer by promoting collagen synthesis [26]. However, researchers now suggest that vitamin C prevents cancer by neutralizing ROS before they can damage DNA and initiate tumor growth. Furthermore, it has been proposed that vitamin C may also act as a pro-oxidant, helping the body's own ROS destroy early-stage tumors [27, 28]. Currently, the recommended dietary allowance (RDA) in many countries ranges from 40–90 mg/day, although the results of various studies suggesting that the protective vitamin C concentrated in plasma for the minimum risk of free radical diseases corresponds to an intake of 124.2 mg/day (in the range of 92–181 mg) [10, 29].

Vitamin C possesses double bonds with an associated electron deficiency, making it highly reactive to free radicals from molecular oxygen. It donates two electrons from C-2 and C-3 double-bonded carbons, resulting in the formation of tricarbonyl ascorbate radical ($AscH^{\bullet}$), which is present in the nonprotonated form, a semidehydroascorbate radical ($Asc^{\bullet-}$). The resulting ascorbate free radicals reduce to a neutral ascorbate molecule (**Figures 2** and **3**, *RIII*). Thus, the oxidation of ascorbate by many ROS is $Asc^{\bullet-}$, a poorly reactive radical that is considered terminal [30–32], and the level of $Asc^{\bullet-}$ radical function is an effective measurement of the degree of oxidative stress in biological systems [33].

In a previous study, we observed that the frequency of micronuclei in polychromatic erythrocytes (MN-PCE) increased with the administration of 40 mg/kg of V_2O_5 through ip [8], consistent with other studies testing soluble vanadium compounds (Na_3VO_4, SVO_5, and NH_4VO_3) [34–36]. However, the *in vivo* administration of vitamin C prior to the V_2O_5 injection decreased MN-PCE formation compared to administering V_2O_5 alone, reducing basal MN-PCE, and presenting the strongest protection against genotoxic damage induced by V_2O_5. This was probably because the vitamin C acted as a potent antioxidant (reducing agent) that scavenged free radicals of reactive oxygen and nitrogen species and prevented them from damaging nucleic acids [27, 37]. Interestingly, the MN-PCE decrease we observed with vitamin C was more effective than with the administration of beverages with high levels of antioxidants such as green tea [38], red wine [39] and their antioxidant components such as polyphenols [40–42].

Despite the important studies on the cytotoxic and anticarcinogenic effects of antioxidants in tumor model systems, it is clear that the molecular mechanisms underlying the benefits of antioxidants in cancer prevention are not yet well understood. Some ascorbyl forms of stearate inhibited cell proliferation by interfering with the cell cycle, reversing the phenotype and inducing apoptosis in human brain tumor glioblastoma (T98G) cells. Therefore, it has been proposed that the chemopreventive properties of antioxidants are related to their ability to target specific cellular signaling pathways that regulate cellular proliferation and apoptosis [43]. This proposal is consistent with our results since the frequencies of apoptotic cells (particularly, late apoptotic cells) indeed increased significantly with the administration of vitamin C, and their administration prior to treatment of V_2O_5 increased them even further [8]. Additionally, other studies have reported that the apoptosis-inducing activity of antioxidants might be synergistically enhanced by a combined treatment with chemopreventive [44] or genotoxic agents [40]. Therefore, it is plausible that enhanced induction of apoptosis following a combined treatment may positively contribute to the elimination of the cells with V_2O_5-induced DNA damage (MN-PCE).

On the other hand, some compounds including vanadium-oxide(V) have been proposed for clinical use as therapeutic drugs for cancer because the intracellular cascade mechanisms may be involved in causing apoptotic cell death. For many decades, vanadium was considered a low-toxicity essential trace element with anticarcinogenic properties [45]. However, important events have taken place since then. In 2006, the International Association for Research on Cancer (IARC) classified vanadium pentoxide (V_2O_5) as a Group 2B substance (possibly carcinogenic to humans) based on results in experimental animals [46]. In 2009, the American Council of Government and Industrial Hygienists (ACGIH) placed V_2O_5 in category A3 (confirmed animal carcinogen with unknown relevance to humans) [47].

The low levels of ROS promoting mRNA formation and encoding proteins known to be regulated by vanadium can induce the activation of transcription factors. In contrast, high levels of ROS are cytotoxic to the cells and trigger apoptotic mechanisms. Therefore, it has been proposed that the cytotoxic effects of vanadium compounds should be used to generate ROS and reactive nitrogen species to combat cancer cell lines [48, 49]. Of all the proposed mechanisms of V(V) toxicity, the induction of oxidative stress is of particular importance for biological systems [50, 51]. As explained above, antioxidants can deactivate highly reactive molecules such as ROS that are generated during various biochemical processes in the cells [3]. As a consequence, substances with antioxidant properties emerge as putative preventatives and co-adjuvants in the treatment of chronic degenerative diseases related to oxidative stress and DNA damage [41]. Additionally, the promising low costs of vanadium-based drugs make it particularly attractive, and the ability to overcome the adverse effects of vanadium compounds during therapeutic action is an urgent and crucial issue for its future use in medicine [49]. Our findings strongly suggest that vitamin C can be used effectively in therapy either alone (antioxidant) or in combination with other agents such as V_2O_5 to reduce their genotoxicity [8].

With regard to Cr(VI) compounds, they have been of particular interest and broadly studied because of their importance in different industrial applications including chrome plating, metallurgy, pigment manufacturing, leather tanning, and wood preservation and, most relevant to this chapter, because they are associated with the induction of cancer [52, 53]. Cr usually exists in various oxidation states, primarily Cr(III) and Cr(VI). The former is an essential micronutrient that plays a key role in protein, sugar, and fat metabolism. The latter is particularly effective in inducing genotoxicity by producing several types of DNA lesions and gene mutations. Some of the major factors that may play a significant role in determining cellular genotoxicity are Cr(VI)-induced DNA-DNA interstrand crosslinks, oxidative DNA damage, and mutations in the tumor suppressor gene p53 [19, 54]. It has been observed that Cr(VI) induces DNA damage through changes in the 8-OH-dG levels in DNA in rats. Furthermore, both endogenic (enzyme system) and exogenic (antioxidant consumption) antioxidant systems might counteract ROS and free radicals. In a recent study, we observed that administration of Cr(VI) increased MN-PCE (genotoxic damage), nonviable cells (cytotoxic damage), and glutathione (GSH) levels (a molecule that intervenes in its reduction to Cr(V), (**Figure 3**, *RII*) and decreased the total levels of antioxidants. Treatments with vitamin C prior to administration of CrO_3 decreased MN frequencies (protection or modulation of genotoxic damage) and nonviable cells (decreased cytotoxic damage). A decrease in the levels of 8-OHdG in CrO_3 group was observed, which could be related to the inhibition of repair mechanisms. However, when the organism was treated with vitamin C, a significant increase in the levels of 8-OHdG was observed, suggesting that it increases DNA repair. Our findings showed a protective effect of vitamin C on genotoxic damage induced by Cr(VI), possibly related to its ROS-suppression properties before the oxidative stress generated by the reduction of Cr(VI) to Cr(III) [55–57]. **Figure 4** summarizes the proposal of the interaction between vitamin C and heavy metals, that is, (1) the free radicals generated by heavy metals can be scavenged by vitamin C inhibiting their genotoxic effects; (2) the repair mechanisms inactivated by heavy metals can be reactivated by vitamin C; and (3) heavy metals induce apoptosis by damaging DNA and vitamin C contributes to this process.

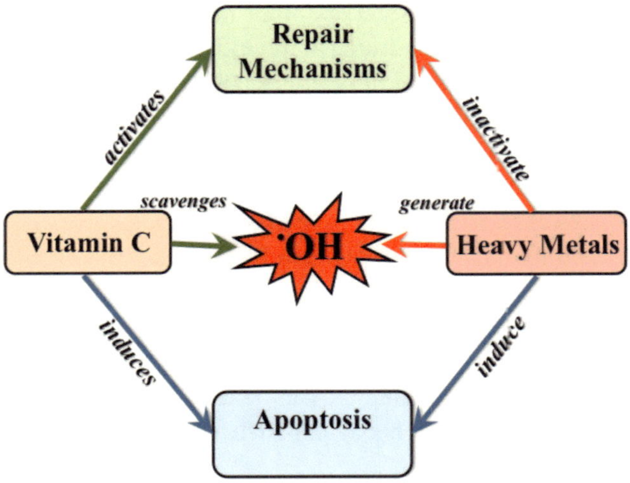

Figure 4. Summary of the interactions between vitamin C and heavy metals.

4. Conclusions

Vitamin C is a potent antioxidant found mainly in fresh fruits and vegetables. It can be readily absorbed and concentrated in tissues and biofluids at a physiologically relevant level, presenting effects in both the aqueous and membrane domains. Furthermore, it plays an essential role in the organism since it scavenges free radicals, chelates redox metals, and regenerates other antioxidants within the "antioxidant network." All these characteristics make the study of the effects of the *in vivo* administration of vitamin C on the genotoxic effects induced by agents associated with oxidative stress particularly important. In this sense, heavy metals such as V(V) and Cr(VI) are of particular relevance since they generate a realistic *in vivo* production of ROS and free radicals due to intracellular reduction. ROS-induced DNA damage involves single- or double-stranded DNA breaks, purine, pyrimidine or deoxyribose modifications, and DNA cross-links. Several studies, including our own results, have solidly concluded that vitamin C does play a significant role in the protection against the genotoxicity caused by metal compounds such as V(V) and Cr(VI). Although the main described mechanism of antioxidants is the scavenging of free radicals, our studies suggest that DNA repair and apoptosis are possible pathways involved in the protection and modulation of DNA. However, it has to be taken into account that under certain conditions (i.e., low concentration *in vitro* of vitamin C and the presence of metal ions), vitamin C can exert a pro-oxidant effect, increasing oxidative damage to DNA. More studies are necessary to fully understand the mechanisms involved in the modulation of and protection against metal-induced genotoxic damage and the adequate doses of vitamin C to stimulate these properties. In addition, it is possible that the impact of vitamin C on DNA damage depends also on both background values of vitamin C within the organism and the level of exposure to xenobiotics or oxidative stress.

http://dx.doi.org/10.5772/intechopen.68686

Conflict of interests

The authors declare that they do not have any competing interests.

Acknowledgements

The authors wish to thank Estefani Y. Hernández-Cruz for his excellent technical assistance. Financial support was obtained from DGAPA-UNAM PAPIIT-IN219216.

Abbreviations

$AscH_2$, $AscH^-$, Asc^{2-}	Forms of ascorbic acid (vitamin C)
$AscH^{\bullet}$	Tricarbonyl ascorbate radical
$Asc^{\bullet-}$	Semidehydroascorbate radical
Cr	Chromium
Cr(VI)	Chromium hexavalent
CrO_3	Chromium trioxide
GSH	Glutathione
H_2O_2	Hydrogen peroxide
ip	Intraperitoneal
MN	Micronucleus
MN-PCE	Micronucleated polychromatic erythrocytes
NADPH	Nicotinamide adenine dinucleotide phosphate
$NAD(P)^+$	Oxidated form of nicotinamide adenine dinucleotide phosphate
Na_3VO_4	Sodium orthovanadate
NH_4VO_3	Ammonium metavanadate
$^{\bullet}OH$	Hydroxyl radical
PCE	Polychromatic erythrocytes
ROS	Reactive oxygen species
SOD	Superoxide dismutase
SVO_5	Vanadyl sulfate
V	Vanadium
$V(IV)–OO^{\bullet}$	Peroxovanadyl radicals
$V(IV)–HO^{\bullet}$	Vanadyl hydroperoxide
V(V)	Vanadium pentavalent
V_2O_5	Vanadium pentoxide
8-OH-dG	8-hydroxydeoxyguanosine

Author details

María del Carmen García-Rodríguez[1]*, Alejandro Gordillo-García[2] and Mario Altamirano-Lozano[1]

*Address all correspondence to: carmen.garcia@unam.mx

1 Research Unit in Genetics and Environmental Toxicology, "Facultad de Estudios Superiores-Zaragoza, UNAM", Mexico City, Mexico

2 Faculty of Arts, University of Leeds, Leeds, UK

References

[1] Seifried HE, Pilch SM. Antioxidants in health and disease. In: Nutrition in the Prevention and Treatment of Disease. 3rd ed. London: Academic Press. 2013. p. 319-339. Available from: http://doi.org/10.1016/B978-0-12-391884-0.00062-7

[2] Combs GF. Vitamin C. In: The Vitamines. 4th ed. London: Elsevier Press. 2012. p. 233-259. Available from: http://doi.org/10.1016/B978-0-12-381980-2.00009-8

[3] Daud ZAM, Ismail A, Sarmadi B. Ascorbic acid: Physiology and health effects. In: Encyclopedia of Food and Health. 1st ed. Oxford: Elsevier Ltd. 2016. p. 266-274. Available from: http://doi.org/10.1016/B978-0-12-384947-2.00045-3

[4] Leonard SS, Harris GK, Shi X. Metal-induced oxidative stress and signal transduction. Free Radical Biology and Medicine. 2004;**37**(12):1921-1942.

[5] Valko M, Rhodes CJ, Moncol J, Izakovic M, Mazur M. Free radicals, metals and antioxidants in oxidative stress-induced cancer. Chemico-Biological Interactions. 2006;**160**(1):1-40. Available from: http://linkinghub.elsevier.com/retrieve/pii/S0009279705004333

[6] Jomova K, Valko M. Advances in metal-induced oxidative stress and human disease. Toxicology. 2011;**283**(2-3):65-87. Available from: http://dx.doi.org/10.1016/j.tox.2011.03.001

[7] Harabawy ASA, Mosleh YYI. The role of vitamins A, C, E and selenium as antioxidants against genotoxicity and cytotoxicity of cadmium, copper, lead and zinc on erythrocytes of Nile tilapia, *Oreochromis niloticus*. Ecotoxicology and Environmental Safety. 2014;**104**(1): 28-35. Available from: http://dx.doi.org/10.1016/j.ecoenv.2014.02.015

[8] García-Rodríguez MC, Hernández-Cortés LM, Altamirano-Lozano MA. In vivo effects of vanadium pentoxide and antioxidants (ascorbic acid and alpha-tocopherol) on apoptotic, cytotoxic, and genotoxic damage in peripheral blood of mice. Oxidative Medicine and Cellular Longevity. 2016;**2016**:1-11. Available from: http://www.hindawi.com/journals/omcl/2016/6797851/

[9] Griffiths HR, Lunec J. Ascorbic acid in the 21st century: More than a simple antioxidant. Environmental Toxicology and Pharmacology. 2001;**10**(4):173-182

[10] Sram RJ, Binkova B, Rossner P. Vitamin C for DNA damage prevention. Mutation Research-Fundamental and Molecular Mechanisms Of Mutagenesis. 2012;**733**(1):39-49

[11] Pisoschi AM, Pop A. The role of antioxidants in the chemistry of oxidative stress: A review. European Journal of Medicinal Chemistry. 2015;**97**:55-74. Available from: http://dx.doi.org/10.1016/j.ejmech.2015.04.040

[12] Fenton HJH. LXXIII—Oxidation of tartaric acid in presence of iron. Journal of the Chemical Society, Transactions. 1894;**65**:899-910. Available from: http://xlink.rsc.org/?DOI=CT8946500899

[13] Marnett LJ. Oxyradicals and DNA damage. Carcinogenesis. 2000;**21**(3):361-370

[14] Cooke MS, Evans MD, Dizdaroglu M, Lunec J. Oxidative DNA damage: Mechanisms, mutation, and disease. FASEB Journal. 2003;**17**(10):1195-1214

[15] Liu KJ, Shi X. In vivo reduction of chromium (VI) and its related free radical generation. Molecular and Cellular Biochemistry. 2001;**222**(1-2):41-47

[16] Shi X, Dalal NS. Hydroxyl radical generation in the Nadh/Microsomal reduction of vanadate. Free Radical Research Communications. 1992;**17**(6):369-376. Available from: http://www.tandfonline.com/doi/full/10.3109/10715769209083141

[17] Liochev SI, Fridovich I. Vanadate-stimulated oxidation of NAD(P)H in the presence of biological membranes and other sources of O_2^-. Archives of Biochemistry and Biophysics. 1990;**279**(1):1-7

[18] Stankiewicz PJ, Stern A, Davison AJ. Oxidation of NADH by vanadium: Kinetics, effects of ligands and role of H_2O_2 or O_2. Archives of Biochemistry and Biophysics. 1991;**287**(1):8-17

[19] Singh J, Carlisle DL, Pritchard DE, Patierno SR. Chromium-induced genotoxicity and apoptosis: Relationship to chromium carcinogenesis (review). Oncology Reports. 1998;**5**(6):1307-1318. Available from: http://www.spandidos-publications.com/10.3892/or.5.6.1307

[20] Maeng S-HH, Chung H-WW, Yu I-JJ, Kim H-YY, Lim C-HH, Kim K-JJ, et al. Changes of 8-OH-dG levels in DNA and its base excision repair activity in rat lungs after inhalation exposure to hexavalent chromium. Mutation Research/Genetic Toxicology and Environmental Mutagenesis. 2003;**539**(1-2):109-116. Available from: http://linkinghub.elsevier.com/retrieve/pii/S1383571803001542

[21] Setyaningsih Y, Husodo AH, Astuti I. Detection of Urinary 8-hydroxydeoxyguanosine (8-OHdG) levels as a biomarker of oxidative DNA damage among home industry workers exposed to chromium. Procedia Environmental Sciences. 2015;**23**:290-296. Available from: http://dx.doi.org/10.1016/j.proenv.2015.01.043

[22] Patlolla AK, Barnes C, Hackett D, Tchounwou PB. Potassium dichromate induced cytotoxicity, genotoxicity and oxidative stress in human liver carcinoma (HepG2) cells. International Journal of Environmental Research and Public Health. 2009;**6**(2):643-653

[23] Rudolf E, Cervinka M. The role of intracellular zinc in chromium(VI)-induced oxidative stress, DNA damage and apoptosis. Chemico-Biological Interactions. 2006;**162**(3):212-227. Available from: http://www.ncbi.nlm.nih.gov/pubmed/16887109

[24] Zhitkovich A. Importance of chromium-DNA adducts in mutagenicity and toxicity of chromium(VI). Chemical Research in Toxicology. 2005;**18**(1):3-11. Available from: http://www.ncbi.nlm.nih.gov/pubmed/15651842

[25] Aditi A, Graham DY. Vitamin c, gastritis, and gastric disease: A historical review and update. Digestive Diseases and Sciences. 2012;**57**(10):2504-2515

[26] Cameron E, Pauling L. Supplemental ascorbate in the supportive treatment of cancer: Prolongation of survival times in terminal human cancer. Proceedings of the National Academy of Sciences of the United States of America. 1976;**73**(10):3685-3689

[27] Du J, Cullen JJ, Buettner GR. Ascorbic acid: Chemistry, biology and the treatment of cancer. Biochimica et Biophysica Acta - Reviews on Cancer. 2012;**1826**(2):443-457. Available from: http://dx.doi.org/10.1016/j.bbcan.2012.06.003

[28] Unlu A, Kirca O, Ozdogan M, Nayır E. High-dose vitamin C and cancer. Journal of Oncological Science. 2016;**1**:10-12

[29] Krajcovicova-Kudlackova M, Dušinská M, Valachovičová M, Blažíček P, Pauková V. Products of DNA, protein and lipid oxidative damage in relation to vitamin C plasma concentration. Physiological Research. 2006;**55**(2):227-231

[30] Atkinson CJ, Nestby R, Ford YY, Dodds PAA. Enhancing beneficial antioxidants in fruits: A plant physiological perspective. BioFactors. 2005;**23**(4):229-234

[31] Williams DJ, Edwards D, Pun S, Chaliha M, Sultanbawa Y. Profiling ellagic acid content: The importance of form and ascorbic acid levels. Food Research International. 2014;**66**:100-106. Available from: http://dx.doi.org/10.1016/j.foodres.2014.09.003

[32] Srdić-Rajić T, Konić Ristić A. Antioxidants: Role on health and prevention. In: Encyclopedia of Food and Health. 1st ed. Oxford: Elsevier Ltd. 2016. p. 227-233. Available from: http://doi.org/10.1016/B978-0-12-384947-2.00038-6

[33] Kašparová S, Brezová V, Valko M, Horecký J, Mlynárik V, Liptaj T, et al. Study of the oxidative stress in a rat model of chronic brain hypoperfusion. Neurochemistry International. 2005;**46**(8):601-611

[34] Ciranni R, Antonetti M, Migliore L. Vanadium salts induce cytogenetic effects in in vivo treated mice. Mutatation Research Toxicology. 1995;**343**(1):53-60

[35] Leopardi P, Villani P, Cordelli E, Siniscalchi E, Veschetti E, Crebelli R. Assessment of the in vivo genotoxicity of vanadate: Analysis of micronuclei and DNA damage induced in mice by oral exposure. Toxicology Letters. 2005;**158**(1):39-49

[36] Rojas-Lemus M, Altamirano-Lozano M, Fortoul TI. Sex differences in blood genotoxic and cytotoxic effects as a consequence of vanadium inhalation: Micronucleus assay evaluation. Journal of Applied Toxicology. 2014;**34**(3):258-264

[37] Crott JW, Fenech M. Effect of vitamin C supplementation on chromosome damage, apoptosis and necrosis ex vivo. Carcinogenesis. 1999;**20**(6):1035-1041. Available from: http://www.ncbi.nlm.nih.gov/pubmed/10357785

[38] García-Rodríguez MC, Vilches-Larrea RE, Nicolás-Mendez T, Altamirano-Lozano MA. El té verde en la quimioprevención in vivo del daño genotóxico inducido por metales cancerígenos (cromo [VI]). Nutricion Hospitalaria. 2012;**27**(4):1204-1212

[39] García-Rodríguez MC, Mateos-Nava RA, Altamirano-Lozano M. Efecto in vivo del vino tinto sin diluir, diluido (75%) y sin alcohol sobre el daño genotóxico inducido por metales pesados con potencial cancerígeno: Cromo [VI]. Nutricion Hospitalaria. 2015;**32**(4):1645-1652

[40] García-Rodríguez MC, Montaño-Rodríguez AR, Altamirano-Lozano MA. Modulation of hexavalent chromium-induced genotoxic damage in peripheral blood of mice by epigallocatechin-3-gallate (EGCG) and its relationship to the apoptotic activity. Journal of Toxicology and Environmental Health Part A. 2016;**79**(1):28-38. Available from: http://www.tandfonline.com/doi/full/10.1080/15287394.2015.1104525

[41] García-Rodríguez MC, Carvente-Juárez MM, Altamirano-Lozano MA. Antigenotoxic and apoptotic activity of green tea polyphenol extracts on hexavalent chromium-induced DNA damage in peripheral blood of CD-1 mice: Analysis with differential acridine orange/ethidium bromide staining. Oxidative Medicine and Cellular Longevity. 2013;**2013**:486419. Available from: http://www.hindawi.com/journals/omcl/2013/486419/

[42] García-Rodríguez MC, Nicolás-Méndez T, Montaño-Rodríguez AR, Altamirano-Lozano MA. Antigenotoxic effects of (−)-epigallocatechin-3-gallate (EGCG), quercetin, and rutin on chromium trioxide-induced micronuclei in the polychromatic erythrocytes of mouse peripheral blood. Journal of Toxicology and Environmental Health A. 2014;**77**(6):324-336. Available from: http://www.ncbi.nlm.nih.gov/pubmed/24593145

[43] Kerr JF, Wyllie AH, Currie AR. Apoptosis: A basic biological phenomenon with wide-ranging implications in tissue kinetics. Journal of Internal Medicine. 1972;**258**(6):479-517

[44] Gao Y, Li W, Jia L, Li B, Chen YC, Tu Y. Enhancement of (−)-epigallocatechin-3-gallate and theaflavin-3-3'-digallate induced apoptosis by ascorbic acid in human lung adenocarcinoma SPC-A-1 cells and esophageal carcinoma Eca-109 cells via MAPK pathways. Biochemical and Biophysical Research Communications. 2013;**438**(2):370-374

[45] Evangelou AM. Vanadium in cancer treatment. Critical Reviews in Oncology/Hematology. 2002;**42**(3):249-265

[46] Altamirano-Lozano M, Beyersmann D, Carter DE, Fowler BA, Fubini B, Kielhorn J, et al. Cobalt in hard metals and cobalt sulfate, gallium arsenide, indium phosphide and vanadium pentoxide. IARC Monographs on the Evaluation of Carcinogenic Risks to Humans. 2006;**86**:1-294.

[47] American Conference of Governmental Industrial Hygienists (ACGIH), "Vanadium pentoxide: Chemical substances 7th edition documentation, 9," in Appendix B: Threshold

Limit Values (TLVs) and Biological Exposure Indices (BEIs). 2012. http://www.nsc.org/facultyportal/Documents/fih-6e-appendixb.pdf

[48] Kioseoglou E, Petanidis S, Gabriel C, Salifoglou A. The chemistry and biology of vanadium compounds in cancer therapeutics. Coordination Chemistry Reviews. 2015;**301-302**:87-105. Available from: http://dx.doi.org/10.1016/j.ccr.2015.03.010

[49] Pessoa JC, Etcheverry S, Gambino D. Vanadium compounds in medicine. Coordination Chemistry Reviews. 2015;**301-302**:24-48. Available from: http://dx.doi.org/10.1016/j.ccr.2014.12.002

[50] Assem FL, Oskarsson A. Vanadium. Handbook on the Toxicology of Metals. 2015;1347-1367. Available from: http://linkinghub.elsevier.com/retrieve/pii/B9780444594532000603

[51] Soriano-Agueda LA, Ortega-Moo C, Garza J, Guevara-García JA, Vargas R. Formation of reactive oxygen species by vanadium complexes. Computational and Theoretical Chemistry. 2016;**1077**:99-105. Available from: http://dx.doi.org/10.1016/j.comptc.2015.11.002

[52] Shi X. Reduction of chromium (VI) and its relationship to carcinogenesis. Journal of Toxicology and Environmental Health, Part B. 1999;**2**(1):87-104. Available from: http://www.tandfonline.com/doi/abs/10.1080/109374099281241

[53] Rowbotham AL, Levy LS, Shuker LK. Critical reviews chromium in the environment: An evaluation of exposure of the UK general population and possible adverse health effects. Journal of Toxicology and Environmental Health, Part B. 2000;**3**(3):145-178

[54] O'Brien TJ, Ceryak S, Patierno SR. Complexities of chromium carcinogenesis: Role of cellular response, repair and recovery mechanisms. Mutation Research/Fundamental and Molecular Mechanisms of Mutagenesis. 2003;**533**(1-2):3-36

[55] García-Rodríguez MC, Serrano-Reyes G, Altamirano-Lozano M. Comparative study in vivo of the genotoxic damage induced by CrO_3 and the effects of the antioxidants: Ascorbic acid, alfa-tocopherol and beta-carotene. Free Radical Biology and Medicine. 2012;**53**:S216. Available from: http://linkinghub.elsevier.com/retrieve/pii/S0891584912009422

[56] García-Rodríguez MC, Serrano-Reyes G, Retana-Ugalde R, Altamirano-Lozano M. Effect of Ascorbic Acid and (-)-Epigallocatechin-3-Gallate on Oxidative Damage Induced by Chromium (VI) in Hsd-ICR Mice. Free Radic Biol Med. 2015;**87**(1):S95. Available from: http://www.sciencedirect.com/science/article/pii/S089158491500862X

[57] Nicolás-Méndez T, Serrano-Reyes G, Pacheco-Martinez M, Altamirano-Lozano M, García-Rodríguez MC. Evaluation of Apoptotic Activity, Micronucleus Induction, and Levels of 8-Hydroxydeoxyguanine and Glutathione in Peripheral Blood of Hsd:ICR Mice Exposed to Hexavalent Chromium Compounds. In: The Toxicologist: Supplement to Toxicological Sciences, 2017;**156**(1):222. Abstract no. 1942

Section 4

Advances in Vitamin C

Chapter 6

Vitamin C Transporter (SVCT2) Distribution in Developing and Adult Brains

Luciano Ferrada, Katterine Salazar and
Francisco Nualart Santander

Additional information is available at the end of the chapter

http://dx.doi.org/10.5772/intechopen.69055

Abstract

Vitamin C is the major antioxidant molecule in the central nervous system (CNS), reaching concentrations of 10 mM in neurons and 400 μM in the cerebrospinal fluid (CSF). Uptake of vitamin C by brain cells is performed through the co-transporter of ascorbic acid and sodium isoform 2, SVCT2, which is expressed in cells from the choroid plexus, neurons, oligodendrocytes, and ependymal cells. SVCT2 expression has also been described in cells at the neurogenic niche, specifically in proliferative type C cells. In this chapter, we will describe recently published studies of SVCT2 expression during brain development and define its polarization in cells from the radial glia (neuronal precursors within the CNS) and vitamin C-mediated effects in regulating genes associated with the maintenance of CNS stem cell pluripotency. We will discuss the differential biological effect that vitamin C generates in neurons versus astrocytes and how the oxidized form of vitamin C, dehydroascorbic acid (DHA), produced by neurons in conditions of oxidative stress must leave this cell to be incorporated by astrocytes. In this context, we will discuss recent literature, which shows that DHA regulates glycolytic metabolism in neurons. In parallel, we will analyze vitamin C recycling by astrocytes, which reduce DHA into ascorbic acid (AA), increasing the antioxidant potential of the brain. Data discussed in this chapter will provide an updated view of SVCT2 distribution in the brain and will also describe how vitamin C recycling participates in normal or pathological brain function.

Keywords: ascorbic acid, dehydroascorbic acid, SVCT2, bystander effect, recycling, brain

1. Introduction

Vitamin C is a small, water-soluble molecule that possesses two dissociable protons with pK values of 4.2 and 11.8. At physiological pH, the reduced form of vitamin C, ascorbic acid

(AA), predominates and is specifically incorporated by neurons through isoform 2 of the cotransporter of sodium and ascorbate (SVCT2) [1]. Once AA loses its protons, it is oxidized into dehydroascorbic acid (DHA), which diffuses among nervous cells through the facilitative glucose transporters, GLUT1 and GLUT3 [2–4]. Most mammals are able to synthesize vitamin C from glucose in the liver; however, primates, including humans, lack the enzyme that catalyzes the last step in vitamin C biosynthesis. Thus, they must obtain vitamin C from the diet [5, 6]. The most relevant function of vitamin C is as an antioxidant agent. Given that the brain has one of the highest metabolic rates of all organs and is under constant oxidative stress, vitamin C is, therefore, critical for the maintenance of brain function and protection of central nervous system (CNS) structures [5].

2. Distribution and function of SVCT2 in the adult brain

Many studies indicate that SVCT2 is a transporter that is preferentially expressed in neurons from different areas in the adult brain, such as the cortex, hippocampus, hypothalamus, and cerebellar precursors [1, 7–11]. SVCT2 mediates uptake of AA, an important molecule for antioxidant defense and general metabolic needs of neurons. Although SVCT2 was originally thought to be a neuronal transporter, there is increasing evidence demonstrating its expression in glial cells in the CNS [4, 7, 8, 10, 12–14]. Indeed, SVCT2 has been reported in cortical microglia, where its function remains unknown [10]. In hypothalamic tanycytes, it may participate in maintaining the high parenchymal concentrations of vitamin C that characterize this region [8]. Furthermore, in choroid plexus cells, it facilitates entry of AA into the brain through the blood-cerebrospinal fluid (CSF) barrier [4, 7, 10, 12]. In Schwann cells, SVCT2 promotes axonal myelination in the peripheral nervous system [14]. Under physiological conditions, SVCT2 protein has not been detected in astrocytes from the gray matter, but its mRNA has been reported in marginal astrocytes in the subpial surface at the entorhinal cortex [13]. In contrast, SVCT2 mRNA is induced in the brains of animals with middle cerebral artery occlusion as an experimental model of ischemia/reperfusion in the brain, suggesting that expression of this transporter is induced under pathological conditions [15]. Moreover, our group has recently defined that SVCT2 protein is induced in reactive astrocytes in various *in vivo* pathological models that generate severe reactive astrogliosis. Specifically, such events involve neuroinflammation generated by intracerebroventricular injection of adenovirus-GFP or the bacterial enzyme, neuraminidase, mechanical damage of the brain cortex evaluated at 5 and 10 days post-injury, and astrogliosis observed in the brain cortex and hippocampus from "Kindled" rats, a widely used model of epilepsy at the mesial temporal lobe [16, 17]. *In vitro* studies support these findings by showing that SVCT2 is induced in astrocytes that have been cultured for long periods and express markers of astrocytic activation [3]. However, we did not observe a positive correlation between SVCT2 expression and the moderate and focal reactive astrogliosis surrounding amyloid plaques in postmortem brain samples from patients with Alzheimer's disease. Altogether, data indicate that SVCT2 is induced in astrocytes during pathophysiological events that produce severe reactive astrogliosis, which may be important to enhance antioxidant defense in order to protect glial cells.

http://dx.doi.org/10.5772/intechopen.69055

3. Expression and polarization of SVCT2 during brain development

Studies published more than a decade ago have unequivocally demonstrated the expression and functionality of SVCT2 in cultured cortical neurons from mouse embryos [18]. Similarly, studies addressing the effect of physiological brain concentrations of AA have shown that this nutrient promotes differentiation of embryonic cortical precursors into neurons and astrocytes [19]. Expression and polarization of SVCT2 have been studied during rat and human embryonic development in the cortex at 9 weeks of gestation [20, 21]. In lissencephaly, SVCT2 is localized in radial glial cells in the ventricular zone, where it is polarized toward the ventricular cavity, and in subapical and intermediate apical progenitors in the subventricular zone [21]. In gyrencephalic brains, SVCT2 localization is also associated with progenitors of the inner and outer areas of the subventricular zone [21]. *In vitro* studies with J1ES cells, a cell model used to evaluate radial glial cell differentiation, indicate that both AA and SVCT2 are key regulators in maintaining the radial phenotype. They also may regulate the pluripotency of neural stem cells, as AA induces the expression of the pluripotency gene, *Nanog*, through activation of the JAK/STAT signaling pathway and inhibition of retinoic acid-dependent neuronal differentiation [21, 22].

During postnatal brain development, levels of SVCT2 mRNA and protein follow an inverse relation with AA concentrations [23]. Thus, high concentrations of AA exist during late embryonic and early postnatal development periods at both the cortex and cerebellum, diminishing later in adulthood; however, levels of SVCT2 mRNA and protein are low during embryonic development but increase with age [23]. By means of confocal and immune-structural studies, our research group has identified SVCT2 in the Golgi apparatus of pyramidal neurons from the deep layers of the brain cortex (layers IV to VI) in early postnatal mice (days 1 and 5 after birth). These findings indicate that SVCT2 is induced during cell arborization and synaptic maturation of deep neurons in the brain cortex. We also identified the expression of a short isoform of SVCT2 (SVCT2sh), which is unable to transport AA and regulate the functional capacity of the active transporter [1]. This finding and its biological consequence in the postnatal brain will be discussed in the following section.

SVCT2 knock-out mice (*SLC23a2* null) have been critical to understanding the physiological importance of SVCT2 in the postnatal brain [24]. Although the intrauterine development of these animals looks normal, homozygote mice that lack SVCT2 present with brain cortex hemorrhage and are unable to expand their lungs, resulting in death a few minutes following birth. Also, SVCT2-null animals have undetectable levels of AA in their brain tissue, supporting the essential role of this transporter in AA entry into cells of the CNS. The defects characterized in SVCT2 knock-out animals suggest that their death was due to CNS problems, possibly generated by a deficit in individual neuronal function, formation of aberrant connections in the cortex and/or other brain areas, or lack of neuronal differentiation in the absence of AA [9]. Further studies have demonstrated that SVCT2 expression is pivotal for neuronal differentiation and maturation, since cultures of hippocampal neurons from knock-out mice have reduced neurite growth, glutamate receptor aggregation, and spontaneous activity [9]. Lentiviral-mediated overexpression of SVCT2 in cells of neuronal linage (N2a) allowed us to demonstrate a three- to four-fold increase in the percentage of cells with increased filopodia numbers and processes that were positive for the dendritic marker, microtubule-associated protein 2 (MAP2) [25]. Furthermore, overexpression of SVCT2 combined with AA treatment promotes phosphorylation of the mitogen-activated protein kinase (MAPK), ERK1/2, a key signaling pathway intermediate that participates in the differentiation and maturation of cortical neurons [25].

The hypothesis that SVCT2 and AA can potentiate neurogenesis during postnatal development has also been validated by Pastor et al. [26], who reported the postnatal expression of SVCT2 in the neurogenic niche from rat and human brains after 1 month of extra-uterine life [26]. This is a restricted area that borders the wall of lateral ventricles in which a stream of progenitor cells that migrate toward the olfactory bulb originate (called the rostral migratory stream [RMS]). SVCT2 is abundant in cells with high proliferative capacity in the neurogenic niche (also known as type C cells) and migratory progenitor cells present at the RMS. In contrast, among the different cell types of the neurogenic niche, SVCT2 is only present at low levels in astrocytes or type B cells and neuroblasts [26]. These findings suggest a role for AA in the maintenance of proliferative cells, which, in turn, is supported by studies indicating that treatment with AA improves the genetic reprogramming of human fibroblasts into induced pluripotent stem cells (iPSC), enhancing the efficiency of the process and allowing the generated iPSC to present similar epigenetic patterns to those present in embryonic stem cells [27]. Moreover, AA supplementation of neurosphere cultures obtained from rat embryonic brain tissue and a teratocarcinoma cell line from P19 mice shows the functionality of SVCT2-mediated uptake of AA as well as neuronal differentiation revealed by the increased expression of the marker, βIII tubulin [26].

In conclusion, SVCT2 is expressed at low levels during embryonic development, and its localization is restricted to the ventricular zone from cerebral ventricles, where the bodies of the radial glia are present. During this period, SVCT2 and AA are key players in regulating pluripotency and the proliferative capacity of neural stem cells. During postnatal development, AA and SVCT2 are important for postmigratory neurons that are beginning the process of arborization and synaptic maturation.

4. Regulation of SVCT2-mediated vitamin C uptake in neurons by shSVCT2

As mentioned in the previous chapter, SVCT2 expression and function are important for the arborization and synaptic maturation of post-mitotic neurons present in the postnatal brain cortex. However, during this period, SVCT2 function is tightly regulated by the co-expression of a shorter isoform of SVCT2, SVCT2sh, which is unable to transport AA. This isoform was initially identified in human fetal brain cDNA [28] and is generated by deletion of an internal sequence of 345 bp in the mRNA, resulting in the translation of a smaller protein completely lacking intracellular domains 5 and 6 and part of domain 4. SVCT2sh was initially thought to be a possible dominant negative of SVCT2 through protein-protein interaction [28]. However, Forster resonance energy transfer (FRET) studies, a decade later using SVCT2-CFP and SVCT2sh-YFP in the neural linage cell line, N2a [1], showed that the proteins interact at both the intracellular space and the cell membrane. The functional consequence of the interaction reduces the affinity of SVCT2 for its substrate, AA, in neuronal cells, suggesting that SVCT2 activity and intracellular concentrations of AA are tightly modulated during postnatal development in order to generate a precise physiological intracellular dose of AA to trigger the neuronal differentiation and synaptic maturation in a particular period of development.

5. Vitamin C recycling

5.1. General aspects

In order to understand the concept of vitamin C recycling, it is necessary to remember that vitamin C is found in two forms: (1) a reduced form, AA, that enters the cells specifically through SVCT transporters and (2) an oxidized form, DHA, that enters the cells through the facilitative glucose transporters, GLUT1, GLUT2, GLUT3, GLUT4, and GLUT8 [29–31]. These two forms of vitamin C represent two different molecules with independent functions and characteristics, although they are interconvertible by means of oxidation-reduction reactions [32]. Because every cell in the body expresses glucose transporters, all cells can uptake vitamin C in the form of DHA [2]. However, for intracellular accumulation of AA and antioxidant protection, the cell needs to reduce DHA into AA, a process that depends on glutathione and NADPH [33–35]. Therefore, the ability of a cell to accumulate AA via reduction of DHA is limited by its antioxidant enzymatic capacity. Interestingly, two different cell populations have been identified: cells with high and those with low antioxidant enzymatic capacity. Cells with low antioxidant enzymatic capacity preferentially use AA for protection against oxidative damage. Once AA accomplishes its antioxidant function, it is oxidized into DHA. Because these cells cannot reduce DHA back into AA, the presence of neighboring cells with high reducing power is crucial. Hence, DHA leaves the cell through glucose transporters and enters neighboring cells that possess an elevated concentration of antioxidant enzymes to reduce DHA into AA. Finally, AA can be accumulated or released into the extracellular space to maintain constant AA concentrations. Therefore, ideal conditions exist *in vivo* to allow coupling between cells in order to efficiently recycle vitamin C.

5.2. Vitamin C recycling between astrocytes and neurons

The idea of vitamin C recycling in the brain originated from the observation that during nutritional deficiency of vitamin C, high AA concentrations in the brain are maintained for a longer period compared to all the other tissues [36, 37]. Thus, efficient mechanisms must exist to maintain AA concentrations in the brain. As mentioned before, in order for vitamin C recycling to occur, a cell with low reducing power that utilizes AA as an antioxidant is required in addition to a cell with high reducing power that is constantly incorporating DHA and releasing AA to the extracellular space. The ideal conditions for vitamin C recycling converge in the CNS, where we find two cell populations that couple in order to recycle vitamin C. On one hand, we find the neuron that is able to intracellularly accumulate up to 10 mM AA but rapidly oxidizes it into DHA, which leaves the cell [4]. Furthermore, the neuron possesses a very limited capacity to reduce DHA back into AA due to low concentration of antioxidant enzymes and its elevated metabolic rate [38]. Accordingly, isolation of neurons from the neighbor cells that recycle DHA compromises their metabolism, resulting in neuronal death [32, 39].

On the other hand, another cell type is found in the CNS, the astrocytes. For a long time, astrocytes were only thought to be scaffold cells in the brain, without any physiological function. However, today we know that astrocytes perform critical physiological activities, such as (1) the uptake of extracellular glutamate for recycling into glutamine to avoid neuronal excitotoxicity; (2) the release of lactate to be used by neurons for energy production; (3) the release of interleukin-6 (IL-6) and tumor growth factor-beta (TGF-beta) neuroprotective cytokines; (4) the synthesis of apolipoprotein E for axonal growth; and (5) the release of trophic factors, such as nerve growth factor (NGF), brain-derived neurotrophic factor (BDNF), and ciliary neurotrophic factor (CNTF), among others [40]. Moreover, astrocytes are required for vitamin C recycling in the CNS due to their ability to uptake DHA from the extracellular space and, in virtue of its elevated reducing power, to efficiently reduce it back into AA [41, 42]. After that, astrocytes release AA to the extracellular space to maintain stable AA concentrations [43]. Besides, astrocytes participate in the neuronal antioxidant defense by releasing AA during pathological conditions that involve changes in the osmotic pressure or ischemic/reperfusion damage [44]. Experiments using neuronal-astrocyte co-cultures have demonstrated that the presence of astrocytes ameliorates neuronal death induced by glutamate, H_2O_2, or DHA *in vitro* [39, 45], suggesting that astrocytes and vitamin C recycling are of vital importance for antioxidant defense in the brain.

6. Relevance of DHA recycling

6.1. Effects of DHA in neurons

As previously described, the neuron utilizes AA as the main antioxidant agent, but it is unable to maintain vitamin C in its reduced form. The effects triggered by the accumulation of DHA in neurons have begun to be elucidated recently. The first studies were focused in determining the effect of the extracellular oxidation of AA over the redox state of the neuron, showing that incubation of brain slices with exogenous AA produces an increase in thiobarbituric acid reactive substances (TBARS) [46–48], an indicator of oxidative damage. However, this effect was prevented by treatment with glutathione, indicating that AA oxidation might induce an increase in reactive oxygen species (ROS). It is widely accepted that AA oxidation generates ROS through the Fenton reaction, which occurs in the presence of metals, such as Fe in acidic pH, producing H_2O_2. AA creates the conditions for the Fenton reaction to occur, so AA might induce an increase in ROS and trigger cell death due to extracellular accumulation of H_2O_2 [49, 50]. Nevertheless, no increase in TBARS occurs when DHA uptake is prevented by inhibition of GLUT1 in brain slices, indicating that the pro-oxidant effect requires DHA uptake by nervous cells and does not depend directly on AA oxidation and accumulation of H_2O_2 [47]. Additionally, as previously described, the pro-oxidant effect

http://dx.doi.org/10.5772/intechopen.69055

of AA disappears when it is co-incubated with glutathione, even when glutathione reduces DHA into AA [33]. Thus, we can conclude that extracellular DHA accumulation due to an oxidative environment stimulates its uptake by neurons, which, in turn, induce oxidative stress because neurons possess low reducing power. Although these results constitute strong evidence suggesting that DHA might be triggering oxidative stress and neuronal death, the ultimate intracellular effect as well as the DHA target has not been identified. Further studies of our research group have shed light on the actual effect of DHA on neurons. Radioactive vitamin C uptake analyses have shown that neurons are able to incorporate both AA and DHA; however, their ability to reduce DHA is limited up to 40% after which they start to oxidize AA back into DHA. Moreover, neurons cannot keep AA in its reduced form for long periods, since it consumes 80% of intracellular AA after 1 h of uptake. However, in the presence of DHA, neurons consume 90% of glutathione in 1 h. Therefore, our group has defined that the neuronal target of DHA is glucose metabolism since incubation with DHA induces a 75% arrest in glycolysis accompanied by increased activity of the pentose phosphate pathway, possibly to regenerate glutathione and avoid oxidative damage [32]. In conclusion, as shown in **Figure 1**, neurons need neighboring cells to recycle vitamin C in order to maintain stable concentrations of AA.

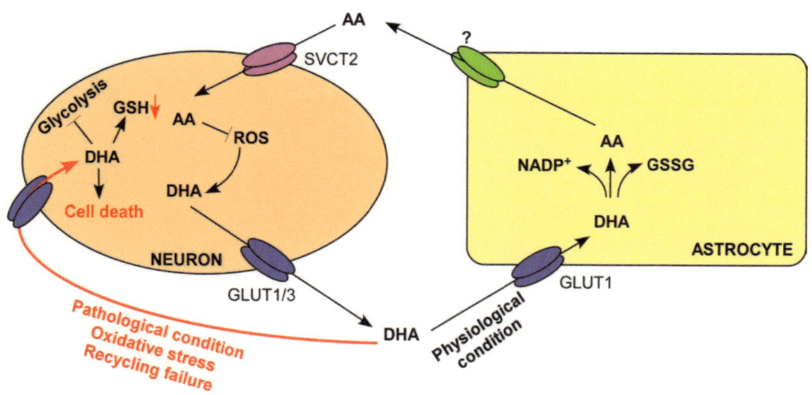

Figure 1. Recycling of vitamin C in physiological and pathophysiological condition. Under physiological conditions, neurons uptake AA through the SVCT2 transporter. AA fulfills its antioxidant function intracellularly, oxidizing to DHA. Following DHA efflux from the neuron through GLUTs 1/3, astrocytes incorporate DHA that was released from the neuron, to reduce it intracellularly to AA at the expense of GSH and NADPH. Finally, the astrocyte releases AA into the extracellular medium to maintain a homeostatically stable concentration of AA. In pathophysiological conditions, vitamin C recycling is deficient; therefore, DHA accumulates in the extracellular space, recycling in an autocrine fashion toward the neuron. Accumulation of DHA in the neuron arrests glycolysis, consumes glutathione, and finally triggers cell death.

6.2. Relevance of astrocytes for DHA recycling

In order to confirm the hypothesis that the presence of vitamin C-recycling cells is necessary, our research group performed studies of cell viability in neurons subjected to oxidative damage induced by H_2O_2 and incubation with DHA. The aim of this experiment was to elucidate the dichotomy existing between the protective and toxic effects that DHA may produce. Given that the literature proposes antioxidant effects for DHA treatment in pathological conditions [51], we incubated neurons with H_2O_2, the major ROS accumulated in the extracellular space during pathological conditions of the brain (e.g., ischemia/reperfusion), and expected a depletion of the antioxidant enzymatic defense of the neuron [52]. In this experimental setting, we incubated neurons with DHA to assess for any antioxidant effect. Interestingly, treatment with H_2O_2 and DHA induced 50 and 70% cell death after 3 and 6 h, respectively, compared with cells that were incubated with H_2O_2 alone [39]. The findings that neurons are inefficient in reducing DHA and that DHA accumulation induces cell death led us to the following question: how do neurons avoid DHA accumulation in physiological and/or pathological conditions? The answer is the presence of another population of brain cells: the astrocytes. To analyze if astrocytes prevent the toxic effects of DHA, neuron-astrocyte co-cultures using the sandwich technique were used to assess any protective effect of astrocytes on neuronal survival in the presence of H_2O_2 and DHA. Neurons were incubated with H_2O_2, and astrocytes were added to the culture just before DHA was added. Notably, neurons co-cultured with astrocytes showed 100% survival after 3 and 6 h of treatment, compared with neurons cultured alone [39], demonstrating that astrocytes are highly efficient in the recycling of DHA from the extracellular medium and that accumulation of DHA due to deficient astrocyte recycling or severe oxidative conditions is toxic for neurons.

7. Conclusions

The relationship between vitamin C and brain pathologies remains a topic of discussion in the current literature, especially given that the oxidized form of vitamin C, DHA, has been largely ignored. DHA is actually a molecule that has great relevance for the treatment of various pathologies, including cancer, although conclusions observed when using DHA should be taken with caution. Indeed, DHA was originally proposed as a molecule that would be beneficial in pathological conditions, such as ischemia and reperfusion [51]. Unfortunately, the research group that proposed DHA as a treatment for ischemia and reperfusion subsequently determined that its administration in primate models did not replicate the results observed in rodents [53]. In these experiments, the researchers did not consider that the observations could be attributed to a failure in vitamin C recycling. Although vitamin C deficiency in the brain has been associated with increased oxidative stress, which could be stimulating ideal conditions to trigger certain pathologies, including Alzheimer's disease [54], DHA has recently been proposed as a molecule that would trigger neuronal death [39]. As shown in **Figure 1**, we propose that CNS pathologies could be associated with reduced astrocytic vitamin C recycling, resulting in the accumulation of DHA in neurons, which would trigger cell death.

Acknowledgements

This work was supported by grants from the Fondo Nacional de Ciencia y Tecnología (FONDECYT 1140477 FN, 11140405 to KS) and Conicyt-PIA CMA BIO BIO ECM-12 to FN.

Author details

Luciano Ferrada, Katterine Salazar and Francisco Nualart Santander*

*Address all correspondence to: frnualart@udec.cl

Center for Advanced Microscopy CMA BIO BIO, Laboratory of Neurobiology and Stem Cells Neuro-CellT, Department of Cellular Biology, Faculty of Biological Sciences, Concepcion University, Concepción, Chile

References

[1] Salazar K, Cerda G, Martinez F, Sarmiento JM, Gonzalez C, Rodriguez F, et al. SVCT2 transporter expression is post-natally induced in cortical neurons and its function is regulated by its short isoform. Journal of Neurochemistry. 2014;**130**:693-706. DOI: 10.1111/jnc.12793

[2] Nualart FJ, Rivas CI, Montecinos VP, Godoy AS, Guaiquil VH, Golde DW, et al. Recycling of vitamin C by a bystander effect. The Journal of Biological Chemistry. 2003;**278**:10128-10133. DOI: 10.1074/jbc.M210686200

[3] Astuya A, Caprile T, Castro M, Salazar K, Garcia Mde L, Reinicke K, et al. Vitamin C uptake and recycling among normal and tumor cells from the central nervous system. Journal of Neuroscience Research. 2005;**79**:146-156. DOI: 10.1002/jnr.20326

[4] Nualart F, Mack L, Garcia A, Cisternas P, Bongarzone ER, Heitzer M, et al. Vitamin C transporters, recycling and the bystander effect in the nervous system: SVCT2 versus gluts. Journal of Stem Cell Research and Therapy. 2014;**4**:209. DOI: 10.4172/2157-7633.1000209

[5] Rice ME. Ascorbate regulation and its neuroprotective role in the brain. Trends in Neurosciences. 2000;**23**:209-216

[6] May JM. Vitamin C transport and its role in the central nervous system. Subcellular Biochemistry. 2012;**56**:85-103. DOI: 10.1007/978-94-007-2199-9_6

[7] Tsukaguchi H, Tokui T, Mackenzie B, Berger UV, Chen XZ, Wang Y, et al. A family of mammalian Na^+-dependent L-ascorbic acid transporters. Nature. 1999;**399**:70-75. DOI: 10.1038/19986

[8] García L, Salazar K, Millan C, Rodriguez F, Montecinos H, Caprile T, et al. Sodium vitamin C cotransporter SVCT2 is expressed in hypothalamic glial cells. Glia. 2005;**50**:32-47. DOI: 10.1002/glia.20133

[9] Qiu S, Li L, Weeber EJ, May JM. Ascorbate transport by primary cultured neurons and its role in neuronal function and protection against excitotoxicity. Journal of Neuroscience Research. 2007;**85**:1046-1056. DOI: 10.1002/jnr.21204

[10] Mun GH, Kim MJ, Lee JH, Kim HJ, Chung YH, Chung YB, et al. Immunohistochemical study of the distribution of sodium-dependent vitamin C transporters in adult rat brain. Journal of Neuroscience Research. 2006;**83**:919-928. DOI: 10.1002/jnr.20751

[11] Oyarce K, Silva-Alvarez C, Ferrada L, Martinez F, Salazar K, Nualart F. SVCT2 is expressed by cerebellar precursor cells, which differentiate into neurons in response to ascorbic acid. Molecular Neurobiology. 2017. DOI: 10.1007/s12035-016-0366-5

[12] Angelow S, Haselbach M, Galla HJ. Functional characterisation of the active ascorbic acid transport into cerebrospinal fluid using primary cultured choroid plexus cells. Brain Research. 2003;**988**:105-113

[13] Nualart F, Salazar K, Oyarce K, Cisternas P, Jara N, Silva-Alvarez C, et al. Typical and atypical stem cells in the brain, vitamin C effect and neuropathology. Biological Research. 2012;**45**:243-256. DOI: 10.4067/S0716-97602012000300006

[14] Gess B, Lohmann C, Halfter H, Young P. Sodium-dependent vitamin C transporter 2 (SVCT2) is necessary for the uptake of L-ascorbic acid into Schwann cells. Glia. 2010;**58**:287-299. DOI: 10.1002/glia.20923

[15] Berger UV, Lu XC, Liu W, Tang Z, Slusher BS, Hediger MA. Effect of middle cerebral artery occlusion on mRNA expression for the sodium-coupled vitamin C transporter SVCT2 in rat brain. Journal of Neurochemistry. 2003;**86**:896-906

[16] Morales JC, Alvarez-Ferradas C, Roncagliolo M, Fuenzalida M, Wellmann M, Nualart FJ, et al. A new rapid kindling variant for induction of cortical epileptogenesis in freely moving rats. Frontiers in cellular neuroscience. 2014;**8**:200. DOI:10.3389/fncel.2014.00200

[17] Cole-Edwards KK, Musto AE, Bazan NG. c-Jun N-terminal kinase activation responses induced by hippocampal kindling are mediated by reactive astrocytes. The Journal of Neuroscience. 2006;**26**:8295-8304. DOI: 10.1523/JNEUROSCI.1986-05.2006

[18] Castro M, Caprile T, Astuya A, Millan C, Reinicke K, Vera JC, et al. High-affinity sodium-vitamin C co-transporters (SVCT) expression in embryonic mouse neurons. Journal of Neurochemistry. 2001;**78**:815-823

[19] Lee JY, Chang MY, Park CH, Kim HY, Kim JH, Son H, et al. Ascorbate-induced differentiation of embryonic cortical precursors into neurons and astrocytes. Journal of Neuroscience Research. 2003;**73**:156-165. DOI: 10.1002/jnr.10647

[20] Caprile T, Salazar K, Astuya A, Cisternas P, Silva-Alvarez C, Montecinos H, et al. The Na^{+}-dependent L-ascorbic acid transporter SVCT2 expressed in brainstem cells, neurons, and neuroblastoma cells is inhibited by flavonoids. Journal of Neurochemistry. 2009;**108**:563-577. DOI: 10.1111/j.1471-4159.2008.05788.x

[21] Silva-Alvarez C, Salazar K, Cisternas P, Martinez F, Liour S, Jara N, et al. Apical polarization of SVCT2 in apical radial glial cells and progenitors during brain development. Molecular Neurobiology. 2016. DOI: 10.1007/s12035-016-0081-2

[22] Wu H, Wu Y, Ai Z, Yang L, Gao Y, Du J, et al. Vitamin C enhances Nanog expression via activation of the JAK/STAT signaling pathway. Stem Cells. 2014;**32**:166-176. DOI: 10.1002/stem.1523

[23] Meredith ME, Harrison FE, May JM. Differential regulation of the ascorbic acid transporter SVCT2 during development and in response to ascorbic acid depletion. Biochemical and Biophysical Research Communications. 2011;**414**:737-742. DOI: 10.1016/j.bbrc.2011.09.146

[24] Sotiriou S, Gispert S, Cheng J, Wang Y, Chen A, Hoogstraten-Miller S, et al. Ascorbic-acid transporter Slc23a1 is essential for vitamin C transport into the brain and for perinatal survival. Nature Medicine. 2002;**8**:514-517. DOI: 10.1038/nm0502-514

[25] Salazar K, Martinez M, Ulloa V, Bertinat R, Martinez F, Jara N, et al. SVCT2 overexpression in neuroblastoma cells induces cellular branching that is associated with ERK signaling. Molecular Neurobiology. 2016;**53**:6668-6679. DOI: 10.1007/s12035-015-9553-z

[26] Pastor P, Cisternas P, Salazar K, Silva-Alvarez C, Oyarce K, Jara N, et al. SVCT2 vitamin C transporter expression in progenitor cells of the postnatal neurogenic niche. Frontiers in Cellular Neuroscience. 2013;**7**:119. DOI: 10.3389/fncel.2013.00119

[27] Esteban MA, Wang T, Qin B, Yang J, Qin D, Cai J, et al. Vitamin C enhances the generation of mouse and human induced pluripotent stem cells. Cell Stem Cell. 2010;**6**:71-79. DOI: 10.1016/j.stem.2009.12.001

[28] Lutsenko EA, Carcamo JM, Golde DW. A human sodium-dependent vitamin C transporter 2 isoform acts as a dominant-negative inhibitor of ascorbic acid transport. Molecular and Cellular Biology. 2004;**24**:3150-3156

[29] Rumsey SC, Daruwala R, Al-Hasani H, Zarnowski MJ, Simpson IA, Levine M. Dehydroascorbic acid transport by GLUT4 in Xenopus oocytes and isolated rat adipocytes. Journal of Biological Chemistry. 2000;**275**:28246-28253. DOI: 10.1074/jbc.M000988200

[30] Rumsey SC, Kwon O, Xu GW, Burant CF, Simpson I, Levine M. Glucose transporter isoforms GLUT1 and GLUT3 transport dehydroascorbic acid. Journal of Biological Chemistry. 1997;**272**:18982-18989

[31] Corpe CP, Eck P, Wang J, Al-Hasani H, Levine M. Intestinal dehydroascorbic acid (DHA) transport mediated by the facilitative sugar transporters, GLUT2 and GLUT8. Journal of Biological Chemistry. 2013;**288**:9092-9101. DOI: 10.1074/jbc.M112.436790

[32] Cisternas P, Silva-Alvarez C, Martinez F, Fernandez E, Ferrada L, Oyarce K, et al. The oxidized form of vitamin C, dehydroascorbic acid, regulates neuronal energy metabolism. Journal of Neurochemistry. 2014;**129**:663-671. DOI: 10.1111/jnc.12663

[33] Li X, Qu ZC, May JM. GSH is required to recycle ascorbic acid in cultured liver cell lines. Antioxidants & Redox Signaling. 2001;**3**:1089-1097. DOI: 10.1089/152308601317203594

[34] May JM, Qu Z, Li X. Requirement for GSH in recycling of ascorbic acid in endothelial cells. Biochemical Pharmacology. 2001;**62**:873-881

[35] Corti A, Casini AF, Pompella A. Cellular pathways for transport and efflux of ascorbate and dehydroascorbate. Archives of Biochemistry and Biophysics. 2010;**500**:107-115. DOI: 10.1016/j.abb.2010.05.014

[36] Hughes RE, Hurley RJ, Jones PR. The retention of ascorbic acid by guinea-pig tissues. The British Journal of Nutrition. 1971;**26**:433-438

[37] Englard S, Seifter S. The biochemical functions of ascorbic acid. Annual Review of Nutrition. 1986;**6**:365-406. DOI: 10.1146/annurev.nu.06.070186.002053

[38] Hediger MA. New view at C. Nature Medicine. 2002;**8**:445-446. DOI: 10.1038/nm0502-445

[39] Garcia-Krauss A, Ferrada L, Astuya A, Salazar K, Cisternas P, Martinez F, et al. Dehydroascorbic acid promotes cell death in neurons under oxidative stress: A protective role for astrocytes. Molecular Neurobiology. 2016;**53**:5847-5863. DOI: 10.1007/s12035-015-9497-3

[40] Allaman I, Belanger M, Magistretti PJ. Astrocyte-neuron metabolic relationships: For better and for worse. Trends in Neurosciences. 2011;**34**:76-87. DOI: 10.1016/j.tins.2010.12.001

[41] Korcok J, Wu F, Tyml K, Hammond RR, Wilson JX. Sepsis inhibits reduction of dehydroascorbic acid and accumulation of ascorbate in astroglial cultures: Intracellular ascorbate depletion increases nitric oxide synthase induction and glutamate uptake inhibition. Journal of Neurochemistry. 2002;**81**:185-193

[42] Daskalopoulos R, Korcok J, Tao L, Wilson JX. Accumulation of intracellular ascorbate from dehydroascorbic acid by astrocytes is decreased after oxidative stress and restored by propofol. Glia. 2002;**39**:124-132. DOI: 10.1002/glia.10099

[43] Wilson JX, Dragan M. Sepsis inhibits recycling and glutamate-stimulated export of ascorbate by astrocytes. Free Radical Biology & Medicine. 2005;**39**:990-998. DOI: 10.1016/j.freeradbiomed.2005.05.020

[44] Wilson JX. Antioxidant defense of the brain: A role for astrocytes. Canadian Journal of Physiology and Pharmacology. 1997;**75**:1149-1163

[45] Kim EJ, Won R, Sohn JH, Chung MA, Nam TS, Lee HJ, et al. Anti-oxidant effect of ascorbic and dehydroascorbic acids in hippocampal slice culture. Biochemical and Biophysical Research Communications. 2008;**366**:8-14. DOI: 10.1016/j.bbrc.2007.11.050

[46] Song JH, Shin SH, Ross GM. Prooxidant effects of ascorbate in rat brain slices. Journal of Neuroscience Research. 1999;**58**:328-336

[47] Song JH, Shin SH, Ross GM. Oxidative stress induced by ascorbate causes neuronal damage in an *in vitro* system. Brain Research. 2001;**895**:66-72

[48] Song JH, Shin SH, Chung IM. Effects of glutamate on dehydroascorbate uptake and its enhanced vulnerability to the peroxidation in cerebral cortical slices. Experimental & Molecular Medicine. 2002;**34**:419-425. DOI: 10.1038/emm.2002.59

[49] Chen Q, Espey MG, Krishna MC, Mitchell JB, Corpe CP, Buettner GR, et al. Pharmacologic ascorbic acid concentrations selectively kill cancer cells: Action as a pro-drug to deliver hydrogen peroxide to tissues. Proceedings of the National Academy of Sciences of the United States of America. 2005;**102**:13604-13609. DOI: 10.1073/pnas.0506390102

[50] Chen Q, Espey MG, Sun AY, Lee JH, Krishna MC, Shacter E, et al. Ascorbate in pharmacologic concentrations selectively generates ascorbate radical and hydrogen peroxide in extracellular fluid *in vivo*. Proceedings of the National Academy of Sciences of the United States of America. 2007;**104**:8749-8754. DOI: 10.1073/pnas.0702854104

[51] Huang J, Agus DB, Winfree CJ, Kiss S, Mack WJ, McTaggart RA, et al. Dehydroascorbic acid, a blood-brain barrier transportable form of vitamin C, mediates potent cerebroprotection in experimental stroke. Proceedings of the National Academy of Sciences of the United States of America. 2001;**98**:11720-11724. DOI: 10.1073/pnas.171325998

[52] Hyslop PA, Zhang Z, Pearson DV, Phebus LA. Measurement of striatal H_2O_2 by microdialysis following global forebrain ischemia and reperfusion in the rat: Correlation with the cytotoxic potential of H_2O_2 *in vitro*. Brain Research. 1995;**671**:181-186

[53] Ducruet AF, Mack WJ, Mocco J, Hoh DJ, Coon AL, D'Ambrosio AL, et al. Preclinical evaluation of postischemic dehydroascorbic acid administration in a large-animal stroke model. Translational Stroke Research. 2011;**2**:399-403. DOI: 10.1007/s12975-011-0084-2

[54] Dixit S, Bernardo A, Walker JM, Kennard JA, Kim GY, Kessler ES, et al. Vitamin C deficiency in the brain impairs cognition, increases amyloid accumulation and deposition, and oxidative stress in APP/PSEN1 and normally aging mice. ACS Chemical Neuroscience. 2015;**6**:570-581. DOI: 10.1021/cn500308h

Chapter 7

Encapsulation of Vitamin C into β-Cyclodextrin for Advanced and Regulatory Release

Subhadeep Saha and Mahendra Nath Roy

Additional information is available at the end of the chapter

http://dx.doi.org/10.5772/intechopen.70035

Abstract

Host-guest inclusion complex (IC) of vitamin C with β-cyclodextrin (β-CD) in aqueous medium has been explored by spectroscopic, physicochemical and calorimetric methods as stabilizer, carrier and regulatory releaser. Job plot has been drawn by UV-visible spectroscopy to confirm the 1:1 stoichiometry of the host-guest assembly. Stereo-chemical nature of the inclusion complex has been explained by two-dimensional (2D) NMR spectroscopy. Surface tension and conductivity studies further support the inclusion process. Association constants for the vitamin C-β-CD inclusion complex have been calculated by UV-visible spectroscopy using both Benesi-Hildebrand method and non-linear programme, while the thermodynamic parameters have been estimated with the help of van't Hoff equation. Isothermal titration calorimetric study has been performed to determine the stoichiometry, association constant and thermodynamic parameters with high accuracy.

Keywords: vitamin C, β-cyclodextrin, inclusion complex

1. Introduction

β-Cyclodextrin (β-CD) is a cyclic oligosaccharide containing seven glucopyranose units, bound by α-(1–4) linkages forming a truncated conical structure [1, 2]. Thus because of its unique structure, i.e. fairly rigid and well-defined hydrophobic cavities and hydrophilic rims having primary and secondary —OH groups (**Figure 1**), it is of particular interest in the modern science [3, 4]. β-CD is used for controlled delivery of organic, inorganic, biological and pharmaceutical molecules due to their ability to form inclusion complexes with diverse guest molecules by encapsulating the non-polar part of the guest into its hydrophobic cavity and stabilizing the polar part by the polar rims [5, 6]. The use of β-CD already has a long history in pharmaceuticals, pesticides, foodstuffs, etc. for the solubility, bioavailability, safety, stability and as a carrier of the guest molecules [7, 8].

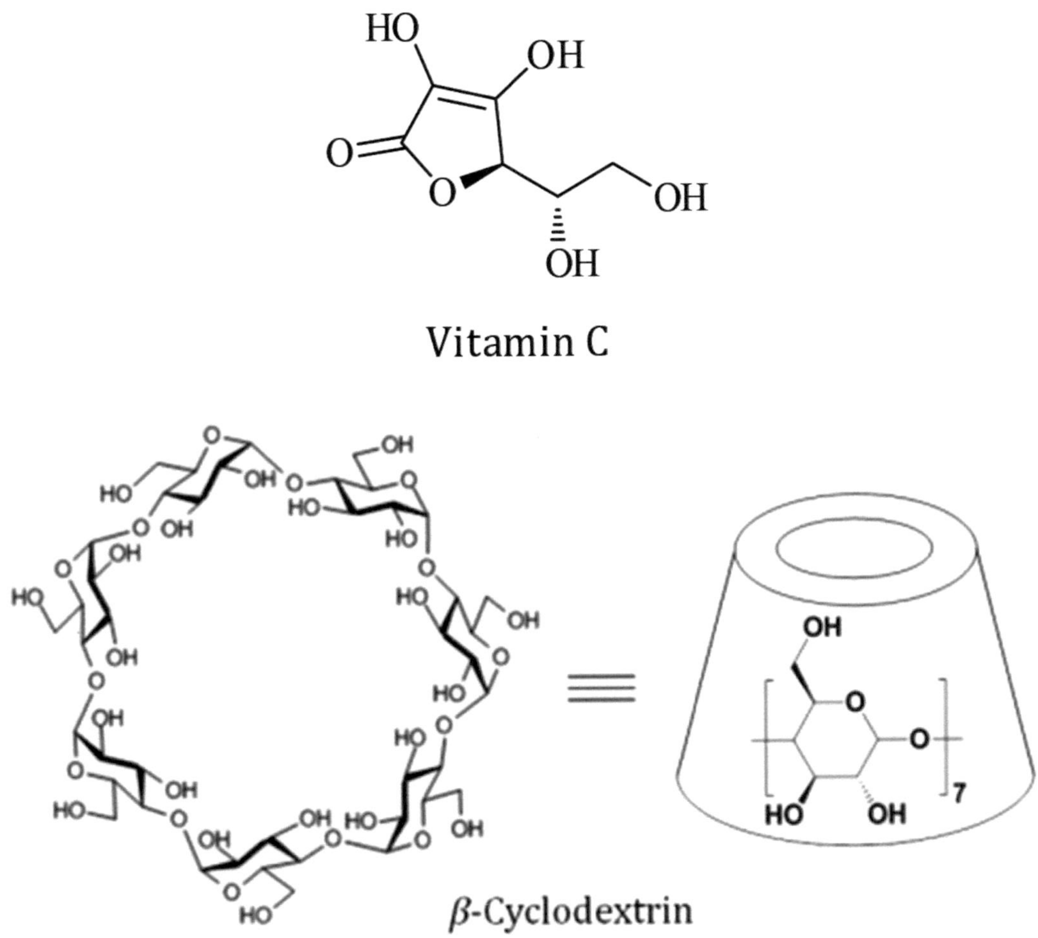

Figure 1. Molecular structure of vitamin C and β-cyclodextrin.

β-CD has been widely employed as not only excellent receptors for molecular recognition but also excellent building blocks to construct functional materials, where they could be applied to construct stimuli-responsive supramolecular materials [9]. Series of external stimuli, e.g. enzyme activation, light, temperature, changes in pH or redox and competitive binding may be employed to operate the release of guest molecules from the inclusion composites [10, 11]. Recently cyclodextrin-modified nanoparticles are of great interest as these supramolecular macrocycles significantly combine and enhance the characteristics of the entities, such as the electronic, conductance, thermal, fluorescence and catalytic properties expanding their potential applications as nanosensors, drug-delivery vehicles and recycling extraction agents [12]. Different sophisticated probes based on semiconductor nanocrystals and other nanoparticles have been designed for this purpose because of their potential applications in the fabrication of molecular switches, molecular machines, supramolecular polymers, chemosensors, transmembrane channels, molecule-based logic gates and other interesting host-guest systems [13–15].

In this article vitamin C (**Figure 1**), is an essential human nutrient with many important functions in biological systems. Scurvy, fatigue, depression and connective tissue defects are the common syndromes caused by deficiency of vitamin C [16, 17]. Thus to protect this important bio-molecule from external effects (e.g. oxidation, structural modification, etc.) and for its regulatory release, it is crucial to investigate whether this molecule can be encapsulated into the β-CD molecule and to explore the thermodynamic aspect of the process. In this present chapter, the formation of host-guest inclusion complex (IC) of the vitamin C with β-CD (the cavity dimension of which is more appropriate than other CDs to encapsulate a great variety of molecules) has been explored particularly towards its formation, stabilization, carrying and controlled release without chemical modification by different dependable methods like two-dimensional rotating-frame nuclear overhauser effect spectroscopy (2D ROESY) NMR, UV-Vis spectroscopy, surface tension (γ), conductivity and isothermal titration calorimetric studies, which primarily focuses on the encapsulation of the bio-molecule into the cavity of β-CD. The stoichiometry, association constants and thermodynamic parameters for the inclusion complex have been determined to communicate a quantitative data regarding the encapsulation of the vitamin by β-CD.

2. Result and discussion

2.1. Job plot reveals the stoichiometry of the host-guest inclusion complex

One of the best methods used to recognize the stoichiometry of the host-guest inclusion complexes is the Job's method, known as the continuous variation method, which has been applied here by using UV-visible spectroscopy [18]. A set of solutions for the vitamin and β-CD was prepared varying the mole fraction of the guest in the range 0–1. Job plot was generated by plotting $\Delta A \times R$ against R, where ΔA is the difference in absorbance of the vitamin without and with β-CD and R = [Vit]/([Vit] + [β-CD]) [19, 20]. Absorbance values were measured at

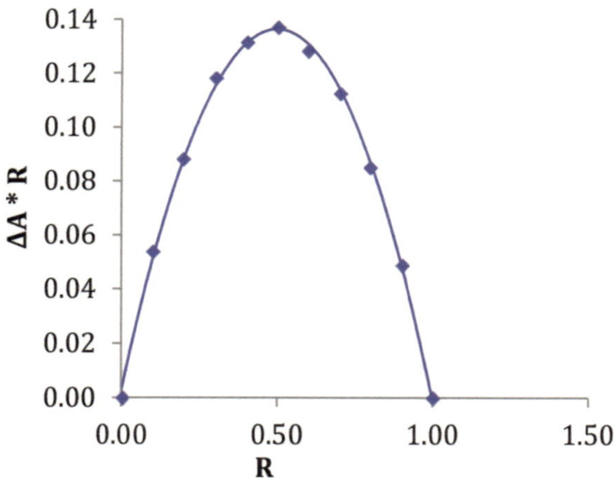

Figure 2. Job plot of vitamin C-β-CD system at 298.15 K. R = [Vit]/([Vit] + [β-CD]), ΔA = absorbance difference of vitamin C without and with β-CD.

λ_{max} for each solution at 298.15 K. The value of R at the maximum deviation gives the stoichiometry of the inclusion complex (IC), i.e. ratio of guest and host is 1:2 if $R = 0.33$; 1:1 if $R = 0.5$; 2:1 if $R = 0.66$, etc. In the present work maxima of the plot was found at $R = 0.5$, which suggest 1:1 stoichiometry of the host-guest inclusion complex (**Figure 2**).

2.2. 2D NMR spectra analysis

Two-dimensional (2D) NMR spectroscopy gives most powerful evidence about the spatial proximity between the host and the guest atoms by observations of the intermolecular dipolar cross-correlations [21, 22]. Any two protons that are located within 0.4 nm in space can

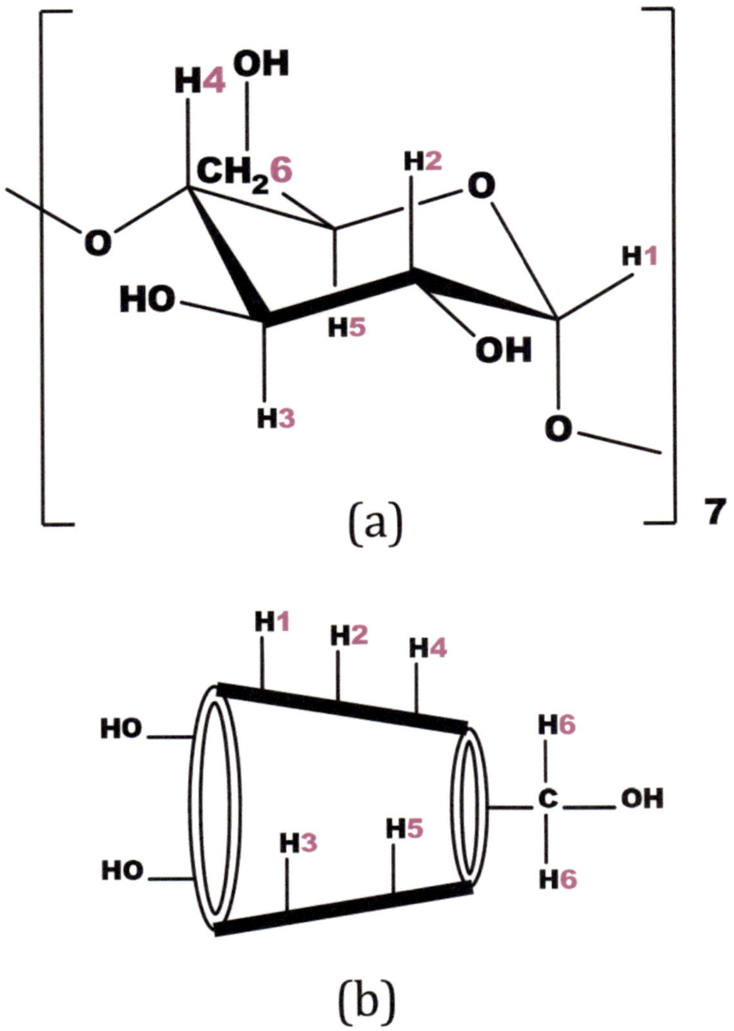

Figure 3. (a) Stereo-chemical configuration of β-cyclodextrin, (b) truncated conical structure of β-cyclodextrin with interior and exterior protons.

produce a nuclear overhauser effect (NOE) cross-correlation in NOE spectroscopy (NOESY) or rotating-frame NOE spectroscopy (ROESY) [23, 24]. In the structure of β-CD the H3 and H5 protons are situated inside the conical cavity, particularly, the H3 are placed near the wider rim while H5 are placed near the narrower rim, the other H1, H2 and H4 protons are located at the exterior of the β-CD molecule (**Figure 3**) [25, 26]. Thus the inclusion phenomenon within the cyclodextrin cavity may be confirmed by the appearance of NOE cross-peaks between the H3 or H5 protons of the host and the protons of the guest identifying their spatial contacts [27, 28]. For this purpose, 2D ROESY has been obtained of the 1:1 molar mixture of vitamin C with β-CD. The ROESY spectra in D_2O shows significant correlations between the H-3, H-5 protons of β-CD and the $C\underline{H}_2$, $C\underline{H}OD$, $C\underline{H}$ protons of vitamin C (**Figure 4**). This result confirms the encapsulation of the vitamin molecule within the cavity

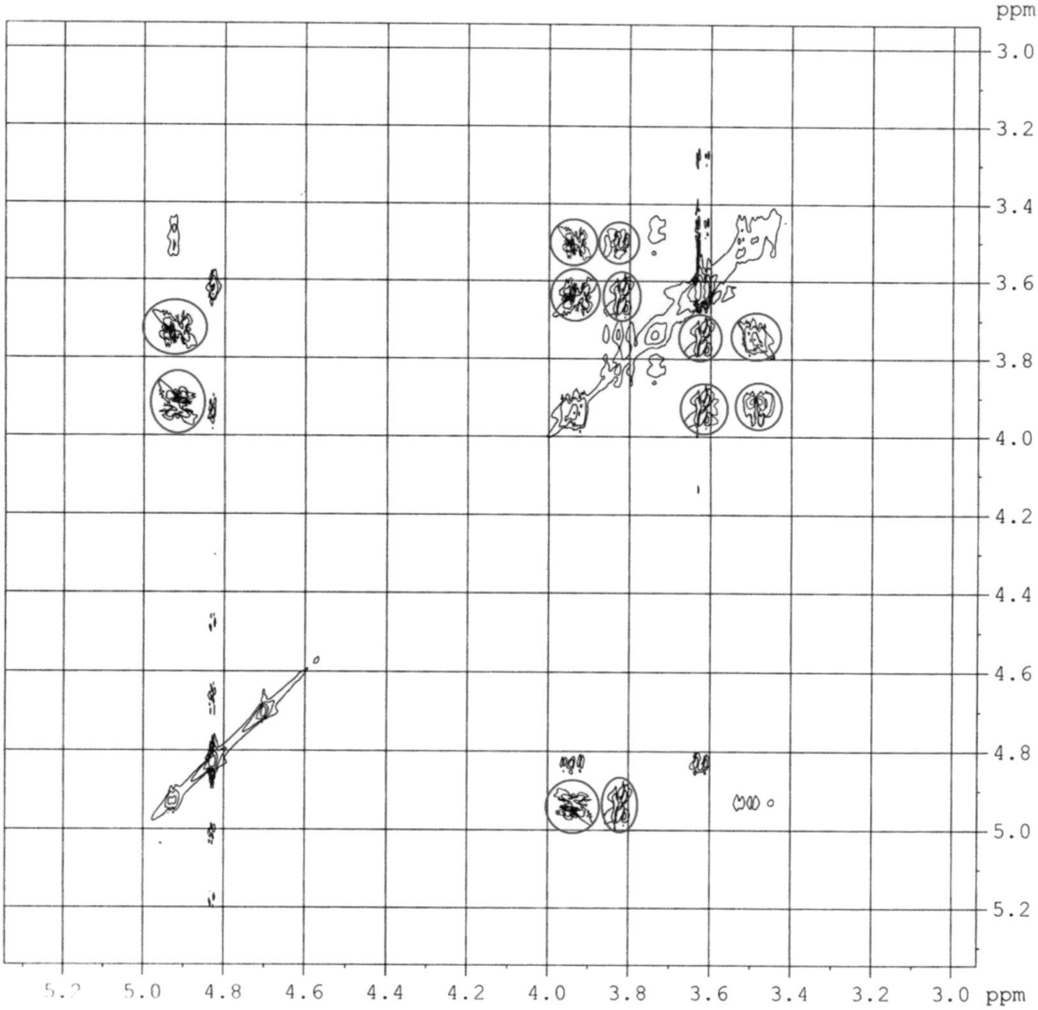

Figure 4. Two-dimensional rotating-frame nuclear overhauser effect spectroscopy (2D ROESY) spectra of 1:1 molar ratio of β-CD and vitamin C in D_2O (correlation signals are marked by circles).

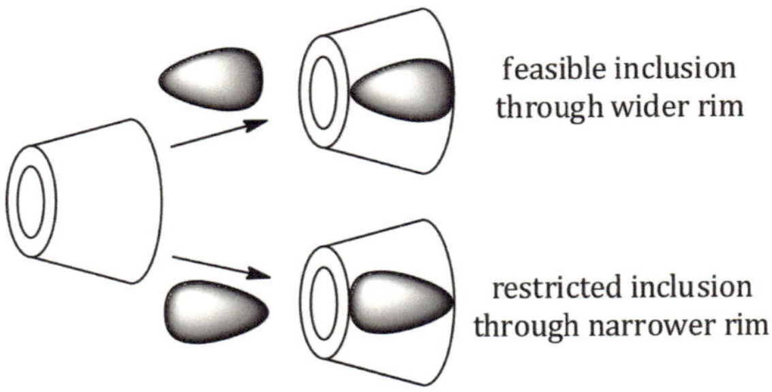

Figure 5. Feasible and restricted inclusion of the guest into the host molecule.

of β-CD. Here in addition, the H6 protons of β-CD were not affected by the inclusion process, which tell that the guest vitamin molecule was included into the β-CD cavity via the wider rim (**Figure 5**) [29].

2.3. Surface tension study elucidates the inclusion as well as stoichiometric ratio of the host and guest

Surface tension (γ) study gives important clue about the formation and the stoichiometry of the host-guest IC [30–32]. Due to ionic interactions there was significant increase in γ of the aqueous solution of vitamin. β-CD, in contrast, because of having hydrophobic outer surface and hydrophilic rims, hardly show any change in γ while dissolved in aqueous medium for a wide range of concentration [32]. In the present study γ of aqueous vitamin has been measured with increasing concentration of β-CD at 298.15 K. The vitamin showed progressively falling trend of γ with increasing concentration of β-CD, may be due to encapsulation of the vitamin molecule from the surface of the solution into the hydrophobic cavity of β-CD forming host-guest inclusion complex (**Figure 6**) [33]. The plot also shows that there is a single discernible break in the curve, which not only points out the formation of IC but also indicates the 1:1 stoichiometric ratio for the IC formed (**Figure 7**) [34, 35]. The value of γ and corresponding concentrations of vitamin and β-CD at break have been listed in **Table 1**, which also indicate that at break point the concentration ratio of host and guest is about 1:1, establishing the formation of 1:1 IC between the studied vitamin and β-CD [8, 36].

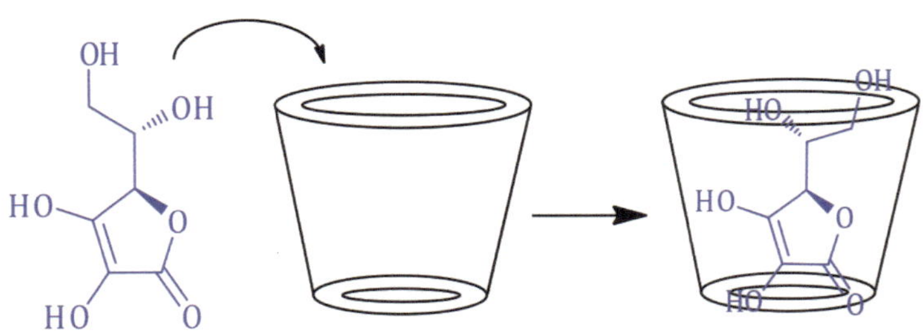

Figure 6. Formation of inclusion complex of vitamin C with β-CD.

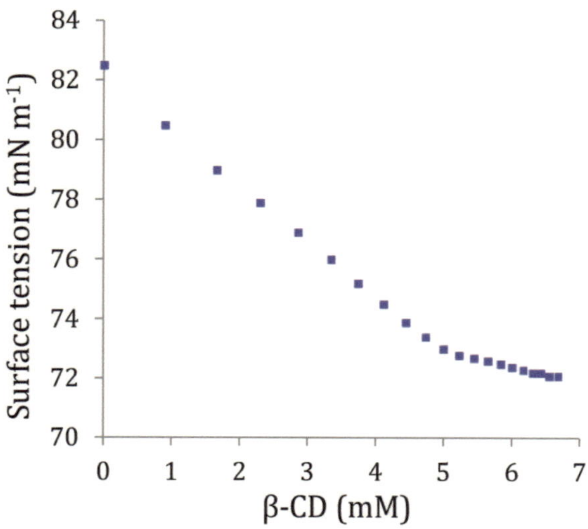

Figure 7. Variation of surface tension of aqueous vitamin C solution with increasing concentration of β-cyclodextrin at 298.15 K.

Concentration of β-CD (mM)	Concentration of vitamin (mM)	γ^a (mN m^{-1})
4.94	5.06	72.98

[a] Standard uncertainties (u): temperature: $u(T) = \pm0.01$ K, surface tension: $u(\gamma) = \pm0.1$ mN m^{-1}.

Table 1. Values of surface tension (γ) at the break point with corresponding concentrations of β-CD and vitamin C at 298.15 K[a].

2.4. Conductivity study demonstrates inclusion process and its stoichiometric ratio

Conductivity (κ) measurement is an important tool to elucidate the inclusion phenomenon in solution phase [30, 32]. It indicates the formation as well as the stoichiometry of the IC formed [37, 38]. In this study, the conductivity of the solution decreases gradually as the vitamin molecules are encapsulated into the cavity of β-CD, i.e. the conductivity of the solution is markedly affected by the inclusion phenomenon. At a certain concentration of β-CD and the vitamin a single break was found in conductivity curve signifying the formation of 1:1 IC (**Figure 8**) [30]. The values of κ and corresponding concentration of the vitamin and β-CD at the break have been listed in **Table 2**, which reveal that the ratio of the concentrations of vitamin C and β-CD at the break point is approximately 1:1, suggesting that vitamin C-cyclodextrin inclusion complex is equimolar, i.e. the host-guest ratio is 1:1 (**Figure 6**).

2.5. Ultraviolet spectroscopy: association constants and thermodynamic parameters

Association constants (K_a) have been calculated for the vitamin-β-CD IC by UV-visible spectroscopy. As the vitamin molecules go from the polar aqueous environment to the apolar

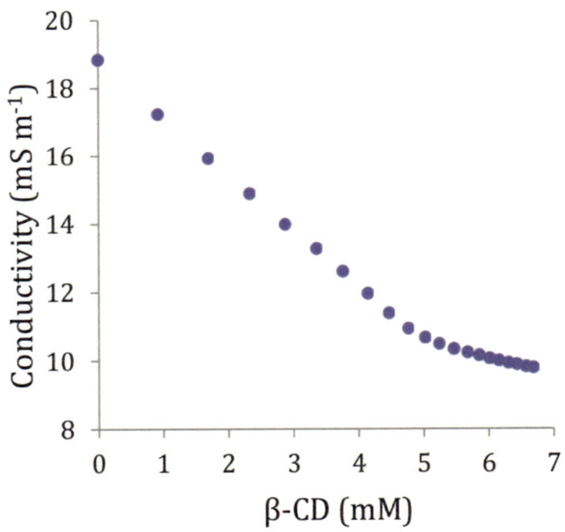

Figure 8. Variation of conductivity of aqueous vitamin C solution with increasing concentration of β-cyclodextrin at 298.15 K.

Concentration of β-CD (mM)	Concentration of vitamin (mM)	κ[a] (mS m^{-1})
4.93	5.07	10.65

[a] Standard uncertainties (u): temperature: u(T) = ±0.01 K, conductivity: $u(\kappa)$ = ±0.001 mS m^{-1}.

Table 2. Values of conductivity (κ) at the break point with corresponding concentrations of β-CD and vitamin C at 298.15 K[a].

cavity of β-CD making the IC, there is a change in molar extinction coefficient ($\Delta\varepsilon$) of the chromophore of the vitamin [39]. The changes in absorbance (ΔA) of vitamin C (261–265 nm) were studied against the concentration of β-CD at different temperatures to determine the association constants (K_a). On the basis of reliable Benesi-Hildebrand method for 1:1 host-guest complex the double reciprocal plots have been drawn using Eq. (1) [20, 40].

$$\frac{1}{\Delta A} = \frac{1}{\Delta\varepsilon[\mathrm{V}]K_\mathrm{a}} \times \frac{1}{[\beta - \mathrm{CD}]} + \frac{1}{\Delta\varepsilon[\mathrm{V}]} \quad (1)$$

The values of K_a for the system were evaluated by dividing the intercept by the slope of the straight line of the double reciprocal plot (**Table 3**) [41, 42].

The thermodynamic parameters can easily be derived basing upon the association constants found at various temperatures by the above method with the help of van't Hoff equation Eq. (2).

$$\ln K_\mathrm{a} = -\frac{\Delta H^\circ}{RT} + \frac{\Delta S^\circ}{R} \quad (2)$$

Temp (K)[a]	$K_a \times 10^{-3}$ (M^{-1})[b]	$\Delta H°$ (kJ mol^{-1})[b]	$\Delta S°$ (J mol^{-1} K^{-1})[b]
288.15	4.19	−21.67	−5.87
293.15	3.58		
298.15	3.10		
303.15	2.68		
308.15	2.33		
313.15	2.03		

[a] Standard uncertainties in temperature u are: $u(T) = \pm 0.01$ K.
[b] Mean errors in $K_a = \pm 0.02 \times 10^{-3}\ M^{-1}$; $\Delta H° = \pm 0.01$ kJ mol^{-1}; $\Delta S° = \pm 0.01$ J mol^{-1} K^{-1}.

Table 3. Association constant (K_a) and thermodynamic parameters $\Delta H°$ and $\Delta S°$ of vitamin C-β-cyclodextrin inclusion complex.

There is a linear relationship between $\ln K_a$ and $1/T$ in the above equation, on the basis of which the thermodynamic parameters $\Delta H°$ and $\Delta S°$ for the formation of IC may be obtained [32, 37, 43].

Association constants (K_a^{ψ}) have also been calculated for the vitamin C-β-CD IC by UV-visible spectroscopy with the help of non-linear programme basing upon the changes in absorbance as a result of encapsulation of the vitamin molecule inside into the apolar cavity of β-CD [44]. The following equilibrium is supposed to exist between the host and the guest for 1:1 IC [1].

$$V_f + CD_f \overset{K_a^{\psi}}{\rightleftharpoons} IC \tag{3}$$

The association constant (K_a^{ψ}) for the formation of IC may be expressed as

$$K_a^{\psi} = \frac{[IC]}{[V]_f[CD]_f} \tag{4}$$

Here, [IC], $[V]_f$ and $[CD]_f$ represent the equilibrium concentration of IC, free vitamin molecule and free CD respectively. According to the binding isotherm, the association constant (K_a^{ψ}) for the formation of IC may be expressed as [45]

$$K_a^{\psi} = \frac{[IC]}{[V]_f[CD]_f} = \frac{(A_{obs} - A_o)}{(A - A_{obs})[CD]_f} \tag{5}$$

where

$$[CD]_f = [CD]_{ad} - \frac{[V]_{ad}(A_{obs} - A_o)}{(A - A_o)} \tag{6}$$

Here, A_o, A_{obs} and A are the absorbance of vitamin molecules at initial state, during addition of CD and final state, respectively. $[V]_{ad}$ and $[CD]_{ad}$ are the concentration of vitamin molecule and the added CD, respectively. Thus, the values of $K_a^{\psi}$for both the systems were evaluated from the binding isotherm by applying non-linear programme (**Table 4**) [7, 46]. The corresponding

Temp (K)[a]	$K_a{}^{\psi} \times 10^{-3}$ (M^{-1})[b]	$\Delta H^{\circ\psi}$ (kJ mol^{-1})[b]	$\Delta S^{\circ\psi}$ (J mol^{-1} K^{-1})[b]
288.15	4.21	−21.83	−6.40
293.15	3.61		
298.15	3.06		
303.15	2.64		
308.15	2.35		
313.15	2.03		

[a] Standard uncertainties in temperature u are: u(T) = ±0.01 K.
[b] Mean errors in $K_a{}^{\psi} = \pm 0.01 \times 10^{-3}$ M^{-1}; $\Delta H^{\circ\psi} = \pm 0.01$ kJ mol^{-1}; $\Delta S^{\circ\psi} = \pm 0.01$ J mol^{-1} K^{-1}.

Table 4. Association constants ($K_a{}^{\psi}$) obtained from non-linear programme and the corresponding thermodynamic parameters $\Delta H^{\circ\psi}$ and $\Delta S^{\circ\psi}$ of vitamin C-β-cyclodextrin inclusion complex.

thermodynamic parameters have been derived basing upon the association constants found from various isotherms by the above method with the help of van't Hoff equation (Eq. (2)) (**Table 4**) [37, 43].

The values of $\Delta H^{\circ}$and $\Delta S^{\circ}$for the formation of IC were found negative suggesting that the inclusion process is exothermic and entropy controlled but not entropy driven (**Table 3**) [37]. These results may be explained on the basis of molecular association that was taking place while the IC was being formed between β-CD and the vitamin. Because of this, there is a drop of entropy, which is unfavourable for the spontaneity of the IC formation. This effect is conquered by higher negative value of ΔH°, making the overall inclusion process thermodynamically favourable.

2.6. Isothermal titration calorimetry: characterization of the complexation

Isothermal titration calorimetry (ITC) is the most sensitive and accurate analytical technique for determination of binding constant and various thermodynamic parameters in host-guest complexation with precise accuracy [47]. It has become an efficient method for direct determination of the thermodynamic parameters rather than using the earlier van't Hoff equation technique [48]. Top of **Figure 9** shows the data obtained from the ITC titration of vitamin C with β-CD in water at 298 K, which describes production of exothermic heat after each injection and the magnitude of the released heat decreases progressively with each injection until complete complexation is achieved. Bottom of **Figure 9** shows the experimental data and the calculated best fit binding curve of vitamin C with β-CD, that provides the stoichiometry (N^{C}), association constant ($K_a{}^{C}$), standard enthalpy ($\Delta H^{\circ C}$) and standard entropy ($\Delta S^{\circ C}$) (**Table 5**). The outcomes of calorimetric study are consistent with those obtained from the analysis of the UV-visible spectroscopic data, however, these values are little different than those obtained by the earlier spectroscopic method studied at a range of temperature, which may be partly illustrated by the fact that the association constants of CD complexes decrease with increasing temperature, on the basis of which the thermodynamic parameters ΔH° and ΔS° were calculated using van't Hoff method. But, in calorimetric study these parameters were determined only at 298 K, thus, the variation of the values of association constants is not

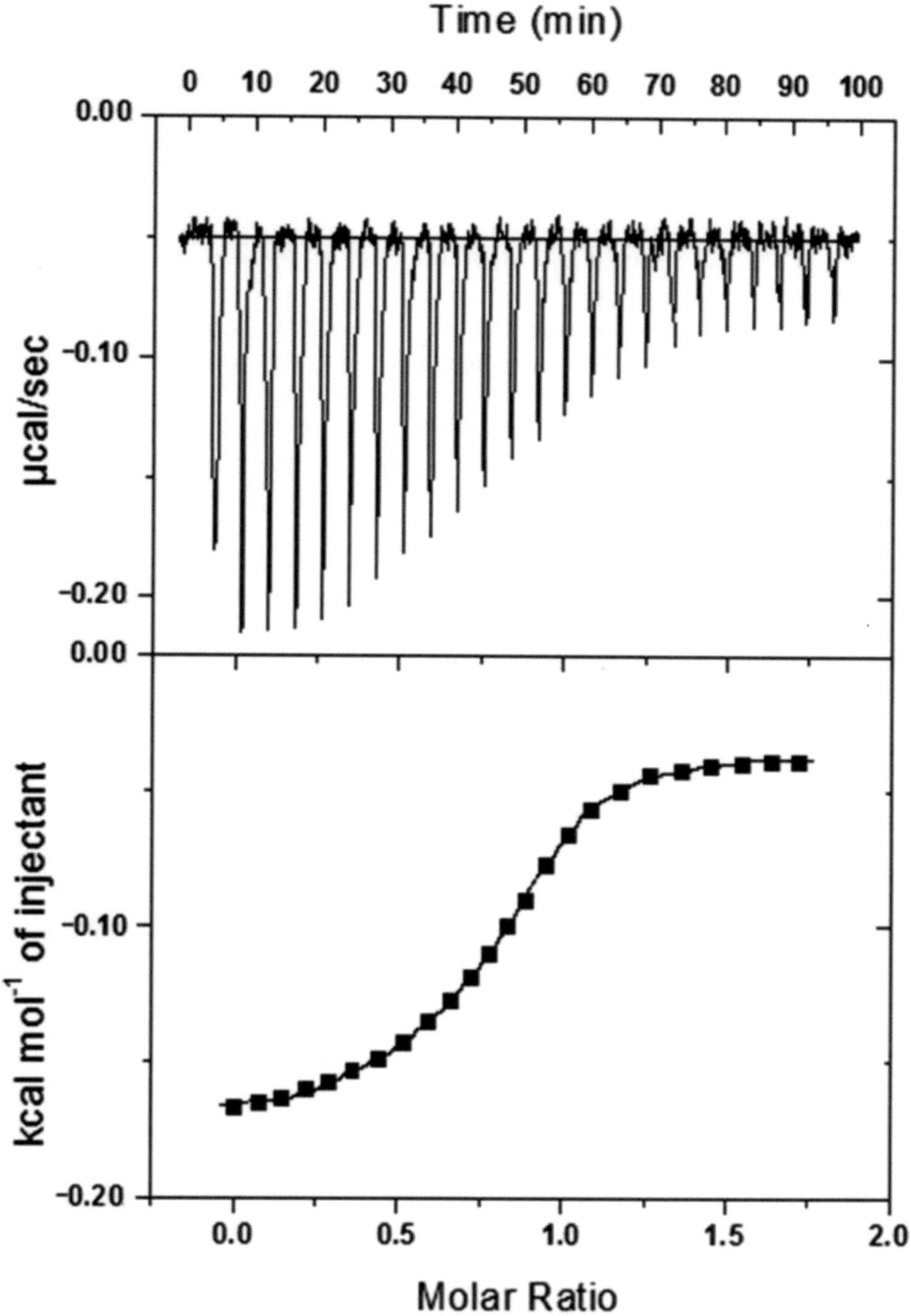

Figure 9. ITC isotherms for the interaction of vitamin C with β-cyclodextrin at 298 K. For each titration, β-cyclodextrin concentration in sample cell was taken as 50 µM and vitamin C concentration in syringe was 500 µM. The top panel represents the raw heats of binding obtained upon titration of vitamin C to β-cyclodextrin. The lower panel is the binding isotherm fitted to the raw data using one site model.

N^C (sites)	$K_a^C \times 10^{-3}$ (M^{-1})	$\Delta H^{\circ C}$ (kJ mol^{-1})	$\Delta S^{\circ C}$ (J mol^{-1} K^{-1})
0.99 ± 0.0111	3.655 ± 0.335	−22.28 ± 1.06	−5.21

Table 5. Stoichiometry (N^C), association constant (K_a^C), standard enthalpy ($\Delta H^{\circ C}$) and standard entropy ($\Delta S^{\circ C}$) of vitamin C-β-cyclodextrin inclusion complex obtained from isothermal titration calorimetric study at 298.15 K.

considered here. The other fact is that in spectroscopic determination, thermodynamic parameters were estimated from association constants, which again were found out on the basis of $\Delta\varepsilon$ of the vitamin, that was due to the changes in the environment around the chromophore, when this goes from the polar aqueous environment to the apolar cavity of β-CD; hence, the changes in enthalpy and entropy described there were exclusively for the formation of IC, not for the other solvent interactions taking place in the medium. But, in calorimetric determination various types of non-covalent forces, like, electrostatic, hydrophobic, van der Waals and H-bonding are involved in the host-guest interaction, thus, thermodynamic parameters represent the overall heat changes resulting from the above interactions [10, 49]. Several mechanisms have been proposed for the complexation, where the most important forces involved are van der Waals and hydrophobic interactions [50]. The binding of vitamin C with β-CD is enthalpy driven as the entropy value of the interaction is not favourable. This indicates electrostatic and hydrophobic interactions play major role in the complexation in this case.

The stoichiometry (N) of the association further suggest that only 1:1 complexation has occurred in the formation of complex of vitamin C with β-CD which is in agreement with the 1:1 complexation revealed from the Job's method.

Formation of the host-guest IC is the dimensional suitability between the two species, which is favoured by the unique cyclodextrin molecule that provides an appropriate condition by encapsulating the apolar part of the guest molecule inside the cavity, as well as stabilizing the polar part by the polar rims [36]. The other driving force for the formation of IC is the release of the water molecules from the hydrophobic cavity into the bulk thereby increasing the entropy of the system [1, 51]. The inclusion of the guest molecule is likely from the wider rim of the β-CD molecule to make maximum contact with the cavity (**Figure 5**), which is also supported by ROESY spectrum. The polar —OH group of the vitamin can also make H-bonds with the —OH groups at both the rims of the β-CD molecule, thereby stabilizing the IC.

3. Experimental

3.1. Materials

Vitamin C and β-cyclodextrin of puriss grade were bought from Sigma-Aldrich, Germany and used as purchased. The mass fraction purity of vitamin C and β-cyclodextrin was ≥0.99 and ≥0.98, respectively.

3.2. Apparatus and procedure

Prior to the start of the experimental work, solubility of β-cyclodextrin and the vitamin has been precisely checked in triply distilled and degassed water (with a specific conductance of 1×10^{-6} S cm^{-1}) and observed that the selected vitamin was freely soluble in all proportion of aqueous β-cyclodextrin. All the stock solutions of the vitamin were prepared by mass (weighed by Mettler Toledo AG-285 with uncertainty 0.0003 g), and then the working solutions were obtained by mass dilution at 298.15 K. Adequate precautions were made to reduce evaporation loss during mixing.

UV-visible spectra were recorded by JASCO V-530 UV/VIS Spectrophotometer, with an uncertainty of wavelength resolution of ±2 nm. The measuring temperature was held constant by an automated digital thermostat.

Two-dimensional (2D) ROESY spectra were recorded in D_2O at 300 MHz using Bruker Avance 300 MHz instrument at 298 K.

The surface tension experiments were done by platinum ring detachment method using a Tensiometer (K9, KRÜSS; Germany) at the experimental temperature. The accuracy of the measurement was within ±0.1 mN m^{-1}. Temperature of the system has been maintained by circulating auto-thermostat water through a double-wall glass vessel containing the solution.

Specific conductance values of the experimental solutions were measured by Mettler Toledo Seven Multi conductivity meter with uncertainty of ±1.0 μS m^{-1}. The measurements were made in an auto-thermostated water bath maintaining the temperature at 298.15 K and using the HPLC grade water with specific conductance of 6.0 μS m^{-1}. The cell was calibrated using a 0.01 M aqueous KCl solution. The uncertainty in temperature was ±0.01 K.

Isothermal titration calorimetry was used to obtain association constant at 298 K using a MicroCal VP-ITC (MicroCal, Inc., Northampton, MA, USA). The thermal equilibration step at 298 K was followed by an initial 120 s delay step and the subsequent 25 injections of each vitamin to β-CD (injection duration of 10 s and spacing of 180 s). Each injection generated a heat-burst curve between micro cal s^{-1} versus time (min). The saturation curve between kcal/mol of injectant versus molar ratio was determined by integration, using Origin 7.0 software (Microcal, Inc.) to give the measure of the heat associated with the injection. The binding affinity and thermodynamic parameters of the binding process were obtained by fitting the integrated heats of binding the isotherm to the one site binding model to give the association constant (K_a^C), stoichiometry (N^C), binding enthalpy ($\Delta H^{\circ C}$) and the entropy ($\Delta S^{\circ C}$).

4. Conclusion

The present study explains that vitamin C forms IC with β-CD in aqueous medium, which can be used as regulatory releaser of the vitamin. Two-dimensional (2D) ROESY NMR study confirms the inclusion phenomenon and its mechanism. Surface tension and conductivity studies also show that the ICs have been formed, the stoichiometry of which were confirmed as 1:1 by Job plots. The association constants and thermodynamic parameters have been estimated for both the ICs by reliable spectroscopic and calorimetric techniques with high accuracy. Thus, this work communicates both qualitative and quantitative idea about the formation of IC of β-CD with vitamin C suggesting its potential applications in pharmaceutical industries and medical sciences.

Author details

Subhadeep Saha and Mahendra Nath Roy*

*Address all correspondence to: mahendraroy2002@yahoo.co.in

Department of Chemistry, University of North Bengal, India

References

[1] Szejtli J. Introduction and general overview of cyclodextrin chemistry. Chemical Reviews. 1998;**98**:1743-1753

[2] Saenger W. Cyclodextrin inclusion compounds in research and industry. Angewandte Chemie International Edition. 1980;**19**:344-362

[3] Bender ML, Komiyama M. Cyclodextrin Chemistry. Berlin: Springer-Verlag; 1978. pp. 10-15

[4] Szejtli J, Osa T. Cyclodextrins. Oxford: Elsevier; 1999. pp. 189-203

[5] Valle EMMD. Cyclodextrins and their uses: A review. Process Biochemistry. 2004;**39**: 1033-1046

[6] Connors KA. The stability of cyclodextrin complexes in solution. Chemical Reviews. 1997;**97**:1325-1357

[7] Kang Y, Guo K, Li BJ, Zhang S. Nanoassemblies driven by cyclodextrin based inclusion complexation. Chemical Communications. 2014;**50**:11083-11092

[8] Gao Y, et al. Inclusion complexes of α-cyclodextrin with ionic liquid surfactants. The Journal of Physical Chemistry B. 2006;**110**:8576-8581

[9] Yu G, Jie Km, Huang F. Supramolecular amphiphiles based on host–guest molecular recognition motifs. Chemical Reviews. 2015;**115**:7240-7303

[10] Zhou Y, et al. Acetylcholine-triggered cargo release from supramolecular nanovalves based on different macrocyclic receptors. Chemistry A European Journal. 2014;**20**:2998-3004

[11] Sun YL, et al. Mechanized silica nanoparticles based on pillar [5]arenes for on-command cargo release. Small. 2013;**9**:3224-3229

[12] Li H, et al. Viologen-mediated assembly of and sensing with carboxylatopillar[5]arene-modified gold nanoparticles. Journal of the American Chemical Society. 2013;**135**;1570-1576

[13] Yu G, Tang G, Huang F. Supramolecular enhancement of aggregation-induced emission and its application in cancer cell imaging. Journal of Materials Chemistry C. 2014;**2**:6609-6617

[14] Yu G, Hua B, Han C. Proton transfer in host−guest complexation between a difunctional pillar[5]arene and alkyldiamines. Organic Letters. 2014;**16**:2486-2489

[15] Angelos S, Yang YW, Khashab NM, Stoddart JF, Zink JI. Dual-controlled nanoparticles exhibiting AND logic. Journal of the American Chemical Society. 2009;**131**:11344-11346

[16] Levine M, et al. Vitamin C pharmacokinetics in healthy volunteers: Evidence for a recommended dietary allowance. Proceedings of the National Academy of Sciences. 1996;**93**:3704-3709

[17] Padayatty SJ, et al. Vitamin C as an antioxidant: Evaluation of its role in disease prevention. Journal of the American College of Nutrition. 2003;**22**:18-35

[18] Job P. Formation and stability of inorganic complexes in solution. Annali di Chimica. 1928;**9**:113-203

[19] Renny JS, Tomasevich LL, Tallmadge EH, Collum DB. Method of continuous variations: Applications of Job plots to the study of molecular associations in organometallic chemistry. Angewandte Chemie International Edition. 2013;**52**:11998-12013

[20] Caso JV, et al. Investigating the inclusion properties of aromatic amino acids complexing beta-cyclodextrins in model peptides. Amino Acids. 2015;**47**:2215-2227

[21] Yang B, Lin J, Chen Y, Liu Y. Artemether/hydroxypropyl-β-cyclodextrin host–guest system: Characterization, phase-solubility and inclusion mode. Bioorganic and Medicinal Chemistry. 2009;**17**:6311-6317

[22] Servais AC, et al. Capillary electrophoretic and nuclear magnetic resonance studies on the opposite affinity pattern of propranolol enantiomers towards various cyclodextrins. Journal of Separation Science. 2010;**33**;1617-1624

[23] Correia I, et al. Study of inclusion complexes of acridine with β- and (2,6-di-*O*-methyl)-β-cyclodextrin by use of solubility diagrams and NMR spectroscopy. Journal of Physical Organic Chemistry. 2002;**15**:647-659

[24] Bednarek E, et al. Complexation of steroid hormones: Prednisolone, ethinyloestradiol and estriol with β-cyclodextrin. An aqueous ^{1}H NMR study. Journal of the Chemical Society Perkin Transactions. 2002;**2**;999-1004

[25] Sindelar V, et al. Supramolecular assembly of 2,7-dimethyldiazapyrenium and cucurbit[8]uril: A new fluorescent host for detection of catechol and dopamine. Chemistry A European Journal. 2005;**11**:7054-7059

[26] Wang T, Wang MD, Ding C, Fu J. Mono-benzimidazole functionalized β-cyclodextrins as supramolecular nanovalves for pH-triggered release of *p*-coumaric acid. Chemical Communications. 2014;**50**:12469-12472

[27] Ishizu T, Hirata C, Yamamoto H, Harano K. Structure and intramolecular flexibility of β-cyclodextrin complex with (−)-epigallocatechin gallate in aqueous solvent. Magnetic Resonance in Chemistry. 2006;**44**;776-783

[28] Kemelbekov U, et al. IR, UV and NMR studies of β-cyclodextrin inclusion complexes of kazcaine and prosidol bases. Journal of Inclusion Phenomena and Macrocyclic Chemistry. 2011;**69**:181-190

[29] Stalin T, Srinivasan K, Sivakumar K, Radhakrishnan S. Preparation and characterizations of solid/aqueous phases inclusion complex of 2,4-dinitroaniline with β-cyclodextrin. Carbohydrate Polymers. 2014;**107**;72-84

[30] Saha S, Ray T, Basak S, Roy MN. NMR, surface tension and conductivity studies to determine the inclusion mechanism: thermodynamics of host–guest inclusion complexes of natural amino acids in aqueous cyclodextrins. New Journal of Chemistry. 2016;**40**: 651-661

[31] Roy A, Saha S, Roy MN. Study to explore host-guest inclusion complexes of cyclodextrins with biologically active molecules in aqueous environment. Fluid Phase Equilibria. 2016;**425**:252-258

[32] Roy MN, Saha S, Kundu M, Saha BC, Barman S. Exploration of inclusion complexes of neurotransmitters with β-cyclodextrin by physicochemical techniques. Chemical Physics Letters. 2016;**655-656**:43-50

[33] Barman S, Ekka D, Saha S, Roy MN. NMR, surface tension and conductance study to investigate host–guest inclusion complexes of three sequential ionic liquids with β-cyclodextrin in aqueous media. Chemical Physics Letters. 2016;**658**;43-50

[34] Pineiro A, et al. On the characterization of host-guest complexes: Surface tension, calorimetry, and molecular dynamics of cyclodextrins with a non-ionic surfactant. The Journal of Physical Chemistry B. 2007;**111**:4383-4392

[35] Gao Y, et al. Preparation and characterization of inclusion complexes of β-cyclodextrin with ionic liquid. Chemistry A European Journal. 2005;**11**:5875-5880

[36] Roy MN, Ekka D, Saha S, Roy MC. Host–guest inclusion complexes of α and β-cyclodextrins with α-amino acids. RSC Advances. 2014;**4**;42383-42390

[37] Roy MN, Saha S, Barman S, Ekka D. Host–guest inclusion complexes of RNA nucleosides inside aqueous cyclodextrins explored by physicochemical and spectroscopic methods. RSC Advances. 2016;**6**:8881-8891

[38] Qian T, Yu C, Wu S, Shen J. Nanoparticles coated polystyrene/reduced graphite oxide microspheres with improved dispersibility and electrical conductivity for dopamine detection. Colloids and Surfaces B: Biointerfaces. 2013;**112**:310-314

[39] Cramer F, Saenger W, Spatz H. Inclusion compounds. The formation of inclusion compounds of α-cyclodextrin in aqueous solutions. Thermodynamics and kinetics. Journal of the American Chemical Society. 1967;**89**:14-20

[40] Benesi HA, Hildebrand JH. A spectrophotometric investigation of the interaction of iodine with aromatic hydrocarbons. Journal of the Chemical Society. 1949;**71**:2703-2707

[41] Dotsikas Y, Kontopanou E, Allagiannis C, Loukas YL. Interaction of 6-*p*-toluidineyl-naphthalene-2-sulphonate with β-cyclodextrin. Journal of Pharmaceutical and Biomedical Analysis. 2000;**23**:997-1003

[42] Roy MN, Roy A, Saha S. Probing inclusion complexes of cyclodextrins with amino acids by physicochemical approach. Carbohydrate Polymers. 2016;**151**:458-466

[43] He Y, Shen X. Interaction between β-cyclodextrin and ionic liquids in aqueous solutions investigated by a competitive method using a substituted 3H-indole probe. Journal of Photochemistry and Photobiology, A: Chemistry. 2008;**197**:253-259

[44] Thordarson P. Determining association constants from titration experiments in supramolecular chemistry. Chemical Society Reviews. 2011;**40**:1305-1323

[45] Spies MA, Schowen RL. The trapping of a spontaneously "flipped-out" base from double helical nucleic acids by host-guest complexation with α-cyclodextrin: The intrinsic base-flipping rate constant for DNA and RNA. Journal of the American Chemical Society. 2002;**124**:14049-14053

[46] Hamdan AA, Bugnon P, Saudan C, Lye PG, Merbach AE. High-pressure studies as a novel approach in determining inclusion mechanisms: Thermodynamics and kinetics of the host-guest interactions for α-cyclodextrin complexes. Journal of the American Chemical Society. 2000;**122**:592-602

[47] Hazra S, Kumar GS. Physicochemical properties of inclusion complexes of sanguinarine with natural cyclodextrins: Spectroscopy, calorimetry and NMR studies. RSC Advances. 2015;**5**:1873-1882

[48] Voskuhl J, et al. Nanodiamonds in sugar rings: An experimental and theoretical investigation of cyclodextrin–nanodiamond inclusion complexes. Organic & Biomolecular Chemistry. 2012;**10**:4524-4530

[49] Kriz Z, Koca J, Imberty A, Charlot A, Velty RA. Investigation of the complexation of (+)-catechin by β-cyclodextrin by a combination of NMR, microcalorimetry and molecular modeling techniques. Organic & Biomolecular Chemistry. 2003;**1**:2590-2595

[50] Ahirwar R, et al. In silico selection of an aptamer to estrogen receptor alpha using computational docking employing estrogen response elements as aptamer-alike molecules. Scientific Reports. 2016;**6**(21285):1-11. DOI: 10.1038/srep21285

[51] Shekaari H, Jebali F. Volumetric and conductometric studies of some amino acids in aqueous ionic liquid, 1-hexyl-3-methylimidazolium chloride solutions at 298.15 K. Physics and Chemistry of Liquids. 2011;**49**:572-587